SpringerBriefs in Electrical and Computer Engineering

SpringerBriefs present concise summaries of cutting-edge research and practical applications across a wide spectrum of fields. Featuring compact volumes of 50 to 125 pages, the series covers a range of content from professional to academic. Typical topics might include: timely report of state-of-the art analytical techniques, a bridge between new research results, as published in journal articles, and a contextual literature review, a snapshot of a hot or emerging topic, an in-depth case study or clinical example and a presentation of core concepts that students must understand in order to make independent contributions.

More information about this series at http://www.springer.com/series/10059

Saleh Faruque

Radio Frequency Cell Site Engineering Made Easy

Saleh Faruque
Department of Electrical Engineering
University of North Dakota
Grand Forks, ND, USA

ISSN 2191-8112 ISSN 2191-8120 (electronic)
SpringerBriefs in Electrical and Computer Engineering
ISBN 978-3-319-99613-4 ISBN 978-3-319-99615-8 (eBook)
https://doi.org/10.1007/978-3-319-99615-8

Library of Congress Control Number: 2018954891

This Springer imprint is published by the registered company Springer Nature Switzerland AG
The registered company address is: Gewerbestrasse 11, 6330 Cham, Switzerland

Preface

Cell site engineering is a system engineering process, which is partly science, partly engineering, and mostly art. This is due to the fact that RF propagation is fuzzy due to uneven terrain, hills, buildings, vegetation, water, etc. For these reasons, it is difficult to accurately predict the RF coverage footprint and cell density. Fortunately, a number of empirical propagation models such as Okumura-Hata model, Cost 200, and Walfisch-Ikegami model are available for initial design. Once the design is complete, we generally collect live air data, perform statistical analysis, and optimize the cell site parameters before offering services to customers.

The list of topics presented in this book is as follows:

- Introduction to cell site engineering
- RF coverage planning
- Choice of multiple access technique
- Frequency planning
- Traffic engineering
- Data collection and optimization

Numerous illustrations are used to bring students up-to-date in key concepts, underlying principles, and practical implementation of cell site engineering.

This book has been primarily designed for electrical engineering students in the area of telecommunications. However, engineers and designers working in the area of wireless communications would also find this book useful. It is assumed that the reader is familiar with the general theory of telecommunications.

In closing, I would like to say a few words about how this book was conceived. It came out of my long industrial and academic career. During my teaching tenure at the University of North Dakota, I developed a number of graduate-level elective courses in the area of telecommunications that combine theory and practice. This book is a collection of my courseware, research activities, and hands-on experience in wireless communications.

I am grateful to UND and the School for the Blind, North Dakota, for affording me this opportunity. This book would never have seen the light of day had UND and the State of North Dakota not provided me with the technology to do so. My heartfelt salute goes out to the dedicated developers of these technologies, who have enabled me and others visually impaired to work comfortably.

I would like to thank my wife, Yasmin, an English literature buff and a writer herself, for being by my side throughout the writing of this book and for patiently proofreading it. My son, Shams, an electrical engineer himself, provided technical support when I needed it. For this, he deserves my heartfelt thanks.
In spite of all this support, there may still be some errors in this book. I hope that my readers forgive me for them. I shall be amply rewarded if they still find this book useful.

Grand Forks, ND, USA
July 11, 2018

Saleh Faruque

Contents

Chapter 1
Introduction

Abstract This chapter provides a brief overview of cellular base station design in a given propagation environment. Beginning with the choice of multiple access technique, we will sip through FCC allocated spectrum, frequency planning, and traffic engineering followed by cell site engineering. Finally, we will see how to obtain live air data and perform data analysis for optimization. Details will be presented in the subsequent chapters.

1.1 Introduction to Cell Site Engineering

Cell site engineering is a system engineering process which is partly science, partly engineering and mostly art [1]. This is due to the fact that RF propagation is fuzzy due to uneven terrain, hills, buildings, vegetation, water, etc. For these reasons, it is difficult to accurately predict the RF coverage footprint and cell density. Fortunately, a number of empirical propagation models, such as Okumura-Hata model, Cost 200, Walfisch-Ikegami model, etc., are available for initial design [2–4]. Once the design is complete, we generally collect live air data, perform statistical analysis, and optimize the cell site parameters before offering services to customers.

In general, cell site engineering involves:

- RF coverage planning
- Choice of technology: access technique, bit rate, and transmission bandwidth
- Frequency planning and optimization
- Traffic engineering
- Cell site integration
- Antenna engineering
- Data collection and optimization

Figure 1.1 shows the basic architecture of land-mobile cellular network for wireline and wireless communications. It comprises the traditional land telephone network and the wireless network. A brief description of this composite network is presented below:

S. Faruque, *Radio Frequency Cell Site Engineering Made Easy*, SpringerBriefs in Electrical and Computer Engineering, https://doi.org/10.1007/978-3-319-99615-8_1

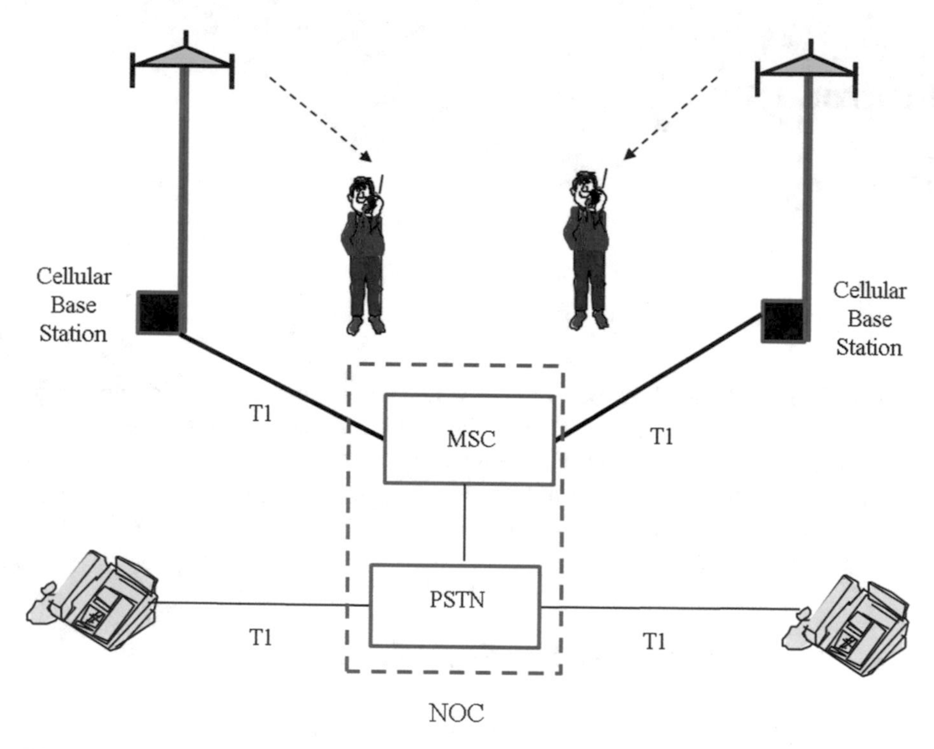

Fig. 1.1 The basic architecture of land-mobile cellular network for wireline and wireless communications

1.2 Wireline Telephone Network

The wireline network is the traditional telephone system in which all telephone subscribers are connected to a central switching network, commonly known as PSTN (public switching telephone network) [5, 6]. It is a computer-based digital switching system, providing the following basic functions:

- Switching
- Billing
- 911 dialing
- 1-800 and 1-900 calling features
- Call waiting, call transfer, conference calling, voice mail, etc.
- Global connectivity
- Interfacing with cellular networks

Tens of thousands of simultaneous calls can be handled by a single PSTN. The composite wired-wireless system, as shown in the figure, is the basis of today's cellular network.

1.3 Wireless Telephone Network

The wireless network is the cellular telephone system in which all cellular telephone subscribers are connected to the base station through wireless links. All base stations are connected to a central switching network through T1 (transmission lines), commonly known as MSC (mobile switching center). The MSC is derived from the PSTN by adding several functions required by the mobile phone system. The function of the MSC is to:

- Provide connectivity between PSTN and cellular base stations by means of trunks (T1 links).
- Facilitate communication between mobile to mobile, mobile to land, land to mobile, and MSC to PSTN.
- Manage, control, and monitor various call processing activities.
- Keep detail record of each call for billing.

1.4 Network Operation Center (NOC)

The Network Operation Center (NOC) contains the PSTN and the MSC. This digital switching network is generally in the same location, supported and maintained by several service personnel. NOC provides connectivity, call processing, billing, etc. In addition, it maintains operation of the network 24/7.

1.5 Cellular Base Station

The cellular base station contains cellular radios, antennas, and supporting equipment, as shown in the figure. The number of radios per base station depends on many factors, such as technology, RF coverage, traffic, etc. For example, the North American AMPS usually requires approximately 48 radios per cell site. Each AMPS radio provides services to one subscriber at a time. On the other hand, the North American TDMA radio can support three subscribers per radio, thus increasing channel capacity threefold. Each base station is connected to the MSC by several T1 links.

Typically, cell phone towers are installed in different propagation environments as follows:

- Every 0.5–1 mile in dense urban environments
- Every 1–3 miles in urban environments
- Every 3–10 miles in suburban environments
- Every 10–30 miles in rural environments

The density of these cell phone towers depends on the technology, antenna pattern, terrain and propagation environment, RF coverage footprints, population density, etc. In a typical metropolitan area, these cellular networks support tens of thousands of users. Notice that the traditional land telephone network is also connected to the cellular network via the NOC (Network Operation Center).

1.6 Base Station Operation

The baseband bipolar signal, received from the T1 link, is converted into NRZ data, demultiplexed, and fed into the radio transmitter for modulation, up-conversion, and amplification. These radio signals are then combined to form a single stream of high-power radio signals, filtered by means of a transmit band-pass filter (duplexer), and transmitted through the main antenna.

On the receive path, the incoming radio signal is filtered (duplexer) and fed to the corresponding radio receivers through the splitter. The function of the splitter is to amplify the incoming signal by means of a low-noise amplifier (LNA), filter, and split into an appropriate number of receive signals, determined by the number of radios on the cell site. An identical receiver path, providing space diversity, recovers the same signal. Both receive signals are then compared for the strongest signal. Finally, the composite multiple signals are placed in the appropriate time slots by means of the multiplexer, converted into bipolar signal, and transmitted to the MSC through T1 links [6].

1.7 Conclusions

This book provides the key concepts, underlying principles, and practical design and operation of cellular networks. Beginning with RF coverage planning, we will review multiple access techniques, FCC allocated spectrum, frequency planning, optimization, traffic engineering, etc. Finally, we will see how to obtain live air data and perform data analysis for cell site optimization.

References

1. S. Faruque, *Cellular Mobile Systems Engineering* (Artech House Inc., Norwood, MA, 1996). ISBN: 0-89006-518-7
2. M. Hata, Empirical formula for propagation loss in land mobile radio services. IEEE Trans. Veh. Technol. **VT-29**, 317–326 (1980)
3. J. Walfisch et al., A theoretical model of UHF propagation in urban environments. IEEE Trans. Antenna Propag. **AP-38**, 1788–1796 (1988)
4. W.C.Y. Lee, *Mobile Cellular Telecommunications Systems* (McGraw-Hill Book Company, New York, 1989)
5. Z. Fluhr, E. Nussbaum, Switching plan for a cellular mobile telephone system. IEEE Trans. Commun. **21**(11), 1281 (1973)
6. V. Hachenburg, B.D. Holm, J.I. Smith, Data signaling functions for a cellular mobile telephone system. IEEE Trans. Veh. Technol. **26**, 82 (1977). https://doi.org/10.1109/T-VT.1977.23660

Chapter 2
RF Coverage Planning

Abstract This chapter presents a brief overview of empirical propagation models and their practical applications along with deployment guidelines. It is shown that all propagation models exhibit equation of straight line within the Fresnel zone break point. A computer-aided design tool is presented as a student project.

2.1 Introduction

RF coverage planning involves site survey, base station location, propagation analysis, RF coverage plots, estimation of cell radii, and the cell count in a given service area. Generally, commercially available computer-aided design tools are used to produce these plots. Figure 2.1 shows a typical RF coverage plot comprising only two cells in an urban environment.

Site survey is an outdoor activity involving:

- Site visits
- Identifying base station location
- Recording lat/long using GPS receiver
- Recording antenna height
- Identifying propagation environment such as dense urban, urban, suburban, rural, etc.

With these design parameters in hand, the next step is to perform propagation analysis by means of a computer-aided design (CAD) tool. These tools begin with empirical propagation models such as Okumura-Hata and Walfisch-Ikegami models, where the geographic information is already built-in. Moreover, these tools require user-defined clutter factors, which are subjective; as a result an error is inevitably present in these tools. Last but not least, these tools are complex and expensive; users require specialized training to use them effectively.

This happens because propagation is fuzzy due to terrain elevation, numerous RF barriers, buildings, trees, water, etc. Building codes also vary from place to place.

S. Faruque, *Radio Frequency Cell Site Engineering Made Easy*, SpringerBriefs in Electrical and Computer Engineering, https://doi.org/10.1007/978-3-319-99615-8_2

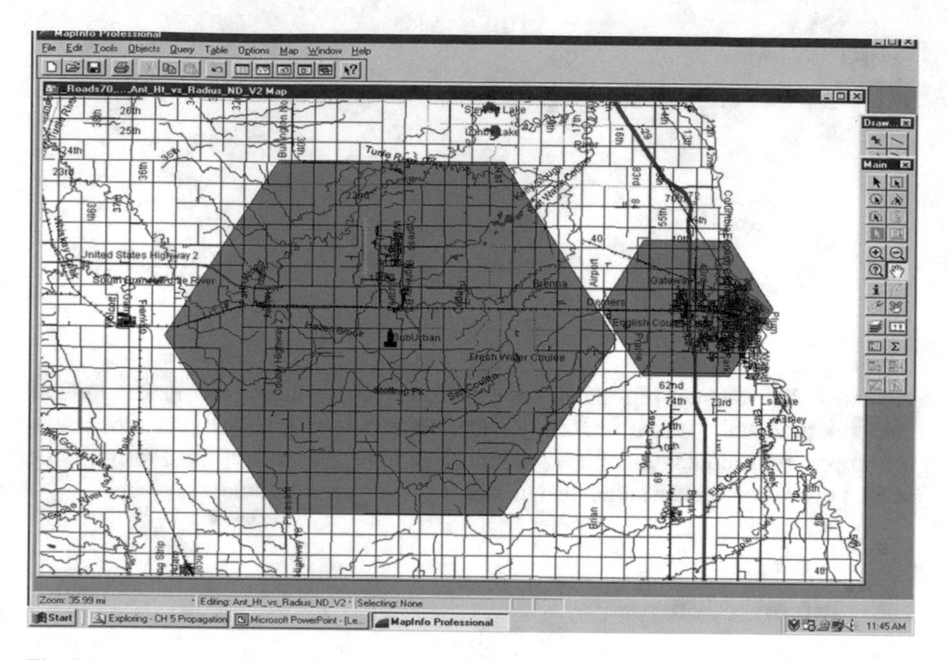

Fig. 2.1 Illustration of a typical RF coverage plot in an urban area

As a result, it is practically impossible to calculate cell radii in terrestrial environment because of multipath propagation components [1]. Multipath propagation arises due to reflection, diffraction, and scattering of radio waves caused by obstructions along the path of transmission. The magnitude of these effects depends on the type and total area of obstruction. For example, a plane surface of vast area will produce maximum reflection, while a sharp object such as a mountain peak or an edge of a building will produce scattering components with minimum effects known as knife-edge effect. These spurious signals have longer path lengths than the direct signal. The associated magnitude and phase differences also vary according to the path length.

Fortunately, there are several empirical propagation prediction models available for designing cellular systems. These prediction models are based on extensive experimental data and statistical analyses, which enable us to compute the received signal level in a given propagation medium. Many commercially available computer-aided prediction tools are based on these models. The usage and accuracy of these prediction models depend on the propagation environment. Among numerous propagation models, the following are the most significant ones, providing the foundation of today's land-mobile communication services [2–4]:

- Okumura-Hata model
- Walfisch-Ikegami model
- Lee model

The usage and accuracy of these prediction models, however, depends chiefly on the propagation environment. For example, the Okumura-Hata model generally provides a good approximation in urban and suburban environments, where the antenna is placed on the roof of the tallest building. On the other hand, the Walfisch-Ikegami model can be applied to dense urban and urban environments, where the antenna height is below the rooftop.

The purpose of this chapter is to review these models and show that all propagation models exhibit equation of straight line having an intercept and a slope. We then classify the propagation environments into four categories:

- Dense urban
- Urban
- Suburban
- Rural

Each propagation environment has a unique path-loss slope. We will then present a PC-based RF planning tool, which has been developed as a student project. It is simple, easy to use, and cost-effective. It uses:

- A GIS software such as MapInfo
- Empirical propagation models
- MS Excel and
- PC

The GPS receiver is used to collect the coordinates of each cell site (lat/long). Once the lat/long of antenna locations are known, the cell radii can be computed in Excel by using any of the existing empirical models such as Okumura-Hata or Walfisch-Ikegami model. Finally, the MapInfo software can be used to import lat/long and the corresponding radius to display the RF coverage plot along with the geographic information such as roads, highways, water, park, buildings, etc. Multiple cell sites can be plotted at the same time by means of this tool.

2.2 Empirical Propagation Models

The purpose of this section is to review empirical models and show that all propagation models exhibit equation of straight line having an intercept and a slope [1]. We then classify the propagation environments into four categories:

- Dense urban
- Urban
- Suburban
- Rural

Each propagation environment has a unique path-loss slope. Let us take a closer look.

2.2.1 *Okumura-Hata Urban and Dense Urban Model*

The Hata model [2] is based on experimental data collected from various urban environments having approximately 16% high-rise buildings. The general path-loss formula of the model is given by:

$$L_p(\mathrm{dB}) = C_o + C_1 + C_2\log(f) - 13.82\log(h_b) - a(h_m) + [44.9 - 6.55\log(h_b)]\mathrm{Log}(d) \tag{2.1}$$

where

L_p = Path loss in dB

$$\begin{aligned} C_o &= 0 \text{ for Urban} \\ &= 3 \text{ dB for Dense Urban} \end{aligned}$$

$$\begin{aligned} C_1 &= 69.55 \text{ for } 150 \text{ MHz} \le f \le 1000 \text{ MHz} \\ &= 46.3 \text{ for } 1500 \text{ MHz} \le f \le 2000 \text{ MHz} \end{aligned}$$

$$\begin{aligned} C_2 &= 26.16 \quad \text{for } 160 \text{ MHz} \le f \le 1000 \text{ MHz} \\ &= 33.9 \quad \text{for } 1600 \text{ MHz} \le f \le 2000 \text{ MHz} \end{aligned}$$

f = Frequency in MHz

h_b = Effective height of the base station in meters [30 m $< h_b <$ 30 m]

$$\begin{aligned} a(h_m) &= \{1.1\log(F) - 0.7\}h_m - \{1.56\log(F) - 0.8\} \text{— for Urban} \\ &= 3.2[\log\{11.75h_m)]^2 - 4.97 \text{— for Dense Urban} \end{aligned}$$

h_m = Mobile antenna height [1 m $< h_m <$ 10 m]

d = Distance between the base station and the mobile (km) [1 km $< d$ $<$ 20 km]

Equation (2.1) may be expressed conveniently as:

$$L_p(\mathrm{dB}) = L_o(\mathrm{dB}) + [44.9 - 6.55\log(h_b)]\log(d) \tag{2.2}$$

or more conveniently as

$$L_p(\mathrm{dB}) = L_o(\mathrm{dB}) + \gamma 10\log(d) \tag{2.3}$$

The above equation exhibits an equation of a straight line, having an intercept L_o and a slope γ:

$$\text{Intercept}: L_o(\mathrm{dB}) = C_o + C_1 + C_2\log(f) - 13.82\log(h_b) - a(h_m) \tag{2.4}$$

$$\text{Slope}: \gamma = [44.9 - 6.55\log(h_b)]/10 \tag{2.5}$$

Note that there are two values for C_o:

- $C_o = 0$ dB for urban and
- $C_o = -3$ dB for dense urban

Also, there are two values for $a(h_m)$:

- $a(h_m) = \{1.1 \log(F) - 0.7\}h_m - \{1.56 \log(F) - 0.8\}$ for urban
- $a(h_m) = 3.2[\log\{11.75h_m)]^2 - 4.97$ for dense urban
- h_m = Mobile antenna height [1 m < h_m < 10 m]

Therefore, the intercept L_o has two values, one for urban and the other for dense urban. On the other hand, the slope γ is the same for both urban and dense urban. It only depends on the base station antenna height. Therefore, we write:

$$\text{For Urban}: L_p(\text{dB}) = L_o(\text{Urban}) + \gamma 10 \log(d) \tag{2.6}$$

$$\text{For Dense Urban}: L_p(\text{dB}) = L_o(\text{Dense Urban}) + \gamma 10 \log(d) \tag{2.7}$$

Figure 2.2 shows two intercept points corresponding to typical urban and dense urban environments, having identical path-loss slopes.

The slope γ is a function of the base station antenna height. Figure 2.3 shows that in a typical urban and dense urban environment, the attenuation slope varies between 3.5 and 4.

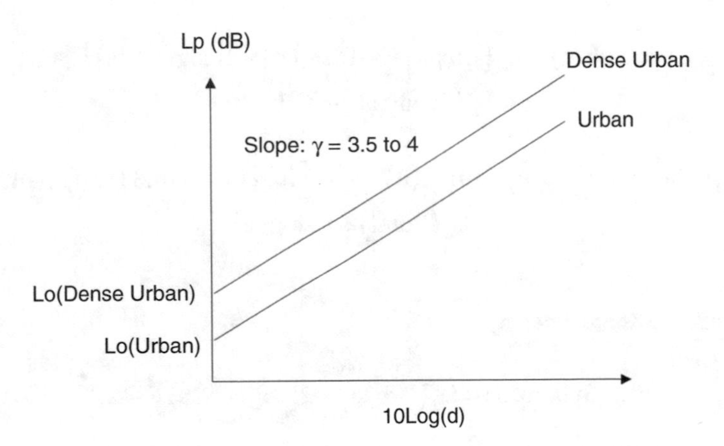

Fig. 2.2 Path-loss characteristics for Okumura-Hata urban and dense urban models

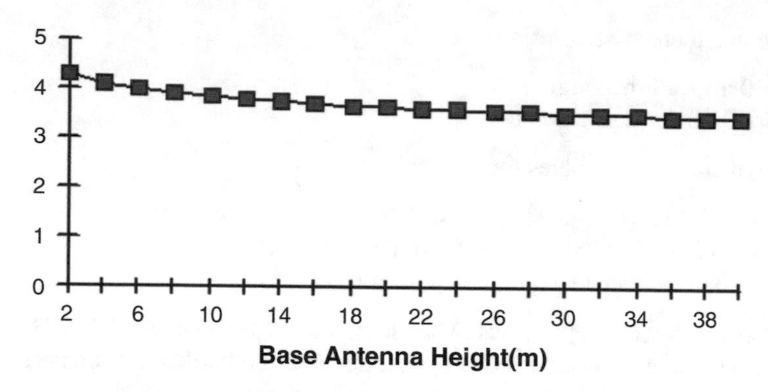

Fig. 2.3 Attenuation slope as a function of base station antenna height in a typical urban and dense urban environment (due to Hata)

2.2.2 Okumura-Hata Suburban and Rural Model

Hata suburban and rural models are based on the urban model with the following corrections:

$$L_p(\text{Suburban}) = L_p(\text{Urban}) - 2\left[\log(f/28)\right]^2 - 2.4 \tag{2.8}$$

$$L_p(\text{rural}) = L_p(\text{Urban}) - 4.78\left[\log(f)\right]^2 + 18.33\log(f) - 40.94 \tag{2.9}$$

Substituting for L_o(Urban) in the above equations, we get:

$$\begin{aligned} L_p(\text{Suburban}) &= L_o(\text{Urban}) + \gamma 10\log(d) - 2\left[\log(f/28)\right]^2 - 2.4 \\ &= L_o(\text{Suburban}) + 10\log(d) \end{aligned} \tag{2.10}$$

$$\begin{aligned} L_p(\text{rural}) &= L_o(\text{Urban}) + 10\log(d) - 4.78\left[\log(f)\right]^2 + 18.33\log(f) - 40.94 \\ &= L_o(\text{Rural}) + 10\log(d) \end{aligned} \tag{2.11}$$

The intercepts are given by:

$$L_o(\text{Suburban}) = L_o(\text{Urban}) - 2\left[\log(f/28)\right]^2 - 2.4 \tag{2.12}$$

$$L_o(\text{Rural}) = L_o(\text{Urban}) - 4.78\left[\log(f)\right]^2 + 18.33\log(f) - 40.94 \tag{2.13}$$

These suburban and rural models also follow the equation of a straight line, having different intercepts, where the path-loss slope remains the same. Figure 2.4 shows the path-loss characteristics for the suburban and the rural models.

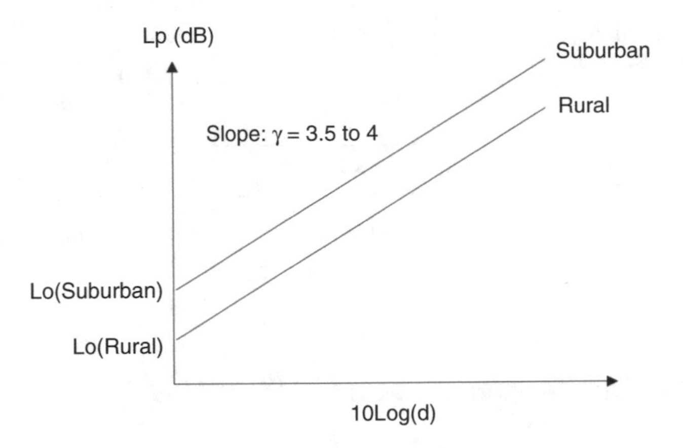

Fig. 2.4 Path-loss characteristics for Okumura-Hata suburban and the rural models

2.2.3 Walfisch-Ikegami Line of Sight (LOS) Model

The Walfisch-Ikegami LOS model [3] is useful for dense urban environments. This model is based on several urban parameters such as building density, average building height, street widths, etc. Antenna height is generally lower than the average building height, so that the signals are guided along the street, simulating an urban canyon-type environment.

For line of sight (LOS) propagation, the path-loss formula is given by:

$$L_p(LOS) = 42.6 + 20\log(f) + 26\log(d) \tag{2.14}$$

where

- f is the frequency in MHz
- d is the distance in km

The above equation can be described by means of the familiar "equation of straight line" as

$$L_p(LOS) = L_o + \gamma 10\log(d) \tag{2.15}$$

where L_o is the intercept and γ is the attenuation slope defined as

$$L_o = 42.6 + 20\log(f) \tag{2.16}$$

$$\gamma = 2.6 \tag{2.17}$$

Such a low attenuation slope in urban environments ($\gamma = 2.6$) is believed to be due to low antenna heights (below the roof top), generating waveguide effects along the street. It follows that if a cell site is located at the intersection of a four-way street, the contour of constant path loss would look like a diamond, as shown in Fig. 2.5. Note that $\gamma = 2$ in free space.

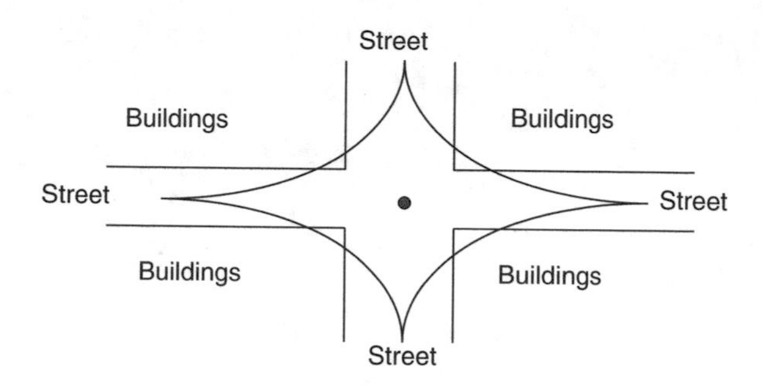

Fig. 2.5 Diamond shape coverage in dense urban canyon

2.2.4 *Walfisch-Ikegami Non-line of Sight Model*

For non-line of sight (NLOS) propagation, the path-loss formula is given by:

$$L_p(\text{NLOS}) = L_p(\text{Free Space}) + L(\text{diff}) + L(\text{mult}) \tag{2.18}$$

Notice that the above equation also exhibits an equation of a straight line, because the free-space path loss exhibits an equation of a straight line and L(mult) and L(diff.) are constants. Let us take a closure look. We define:

- L_p(Free Space) = 32.5 + 20 log(f) + 20 log(d)
- f, d = Frequency and distance, respectively
- L(diff.) = Rooftop diffraction loss
- L(mult) = Multiple diffraction loss due to surrounding buildings

The rooftop diffraction loss is characterized as:

$$L(\text{diff.}) = -16.9 - 10\log(\Delta W) + 10\log(f) + 20\log(\Delta h_m) + L(\phi) \tag{2.19}$$

The parameters in the above equation are defines as:

- ΔW = Distance between the street mobile and the building
- h_m = Mobile antenna height
- $\Delta h_m = h_{roof} - h_m$
- L(ø) = Loss due to elevation angle

The above parameters are constants after the antenna is installed. Therefore, L(diff.) is constant.

Multiple diffraction and scattering components are characterized by following equation:

$$L(\text{mult}) = k_o + k_a + k_d \cdot \log(d) + k_f \cdot \log(f) - 9\log(W) \tag{2.20}$$

where

- $k_o = -18\log(1 + \Delta h_b)$
- $k_a = 54 - 0.8(\Delta h_b) \quad d \geq 0.5 \text{ km}$
  $= 54 - 0.8(\Delta h_b) \quad d \leq 0.5 \text{ km}$
- $k_d = 18 - 16(\Delta h_b / h_{roof})$
- $k_f = -4 + 0.7[(f/925) - 1] \quad \text{for suburban}$
  $= -4 + 1.5[(f/925) - 1] \quad \text{for urban}$
- W = Street width
- h_b = Base station antenna height
- h_{roof} = Average height of surrounding small buildings ($h_{roof} < h_b$)
- $\Delta h_b = h_b - h_{roof}$

The above parameters are also constants after the antenna is installed.

We assumed that the base station antenna height is lower than tall buildings but higher than small buildings. Combining above equations we obtain:

$$
\begin{aligned}
L_p(\text{NLOS}) &= L_o + (20 + k_d)\log(d) \\
&= L_o + \gamma\ 10\log(d)
\end{aligned}
\tag{2.21}
$$

The arbitrary constants are lumped together to obtain:

$$
\begin{aligned}
L_o &= 32.4 + (30 + k_f)\log(f) - 16.9 - 10\log(w) + 20\log(\Delta h_m) \\
&\quad + L(\phi) + k_o + k_a - 9\log(W) \\
\gamma &= (20 + k_d)/10
\end{aligned}
\tag{2.22}
$$

Once again, the NLOS path-loss characteristics also exhibit an equation of straight line with L_o as the intercept and γ as the slope.

The diffraction constant k_d depends on surrounding building heights, which vary from one urban environment to another, yielding a diffraction constant of a few meters to tens of meters. Typical attenuation slopes in these environments range from $\gamma = 2$ for $\Delta h_b/h_{roof} = 1.2$ to $\gamma = 3.8$ for $\Delta h_b/h_{roof} = 0$. This is shown in Fig. 2.6.

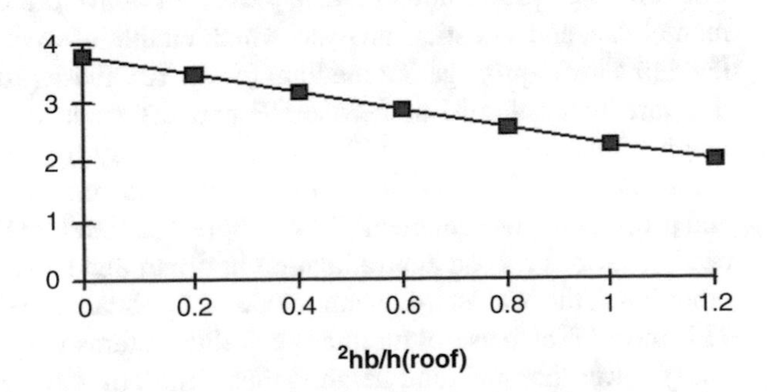

Fig. 2.6 Attenuation slope as a function of base station antenna height in a typical dense urban environment (due to Walfisch-Ikegami)

2.2.5 *Lee Model*

In order to accommodate terrestrial factors, C.Y. Lee has developed a simplified formula, given by [4]:

$$L_p = 129.45 + 38.4\log(d) - 20\log(h_b) \quad \text{for 900 MHz Cellular} \tag{2.23}$$

$$L_p = 132.45 + 38.4\log(d) - 20\log(h_b) \quad \text{for 1900 MHz Cellular (PCS)} \tag{2.24}$$

where

- L_p = Path loss in dB
- d = Distance in km
- h_b = Base station antenna height in meters
- C = 129.45 dB is the average loss in typical urban environment (f = 900 MHz)
- C = 132.45 dB is the average loss in typical urban environment (f = 1900 MHz)

The above equations can be conveniently written as follows:

$$L_p(900) = C_1 + 38.4\log(d) \tag{2.25}$$

$$L_p(1900) = C_2 + 38.4\log(d) \tag{2.26}$$

Once again, we see that Lee model also exhibits an equation of a straight line, where C_1 and C_2 are intercepts and γ is the slope:

- C_1 = 129.45 – 20 log(H_b) (H_b = Base station antenna height)
- C_2 = 132.45 – 20 log(H_b) (H_b = Base station antenna height)
- γ = 3.84 (slope)

2.3 RF Engineering Guidelines

The empirical propagation models presented above are based on extensive experimental data and statistical analysis which enable us to compute the received signal level in a given propagation medium [5–7]. Yet, these propagation models, in practice, are fuzzy due to numerous RF barriers such as uneven terrain, buildings' heights, hills, trees, etc.; building codes also vary from place to place. As a result, the accuracy of these prediction models depends on the frequency, antenna height, and propagation environment. For example, the standard Okumura-Hata model generally provides a good approximation in urban and suburban environments. On the other hand, the Walfisch-Ikegami model is applicable to dense urban environments. This model is also useful for micro-cellular systems where antenna heights are generally lower than building heights, thus simulating an urban canyon environment. Lee model can also be used for 900 MHz macro cell and 1900 MHz PCS system.

Table 2.1 RF deployment guidelines

Propagation environments	Typical path-loss slope (γ)	Propagation models
Dense urban: • High-rise buildings "canyon" channel propagation • Antennas above the rooftop, causing multiple diffractions • Antennas below the rooftop, causing multiple reflections	4	Walfisch-Ikegami model
Urban: • Mixture of various building heights and open areas		Okumura-Hata model
Suburban: • Residential areas • Open fields	3	Okumura-Hata model
Rural: • Farm areas • Highways	2.5	Okumura-Hata model
Free space: • Outer space • Terrestrial environments: distance within the Fresnel zone break point. All propagation environments. Depends on frequency and antenna height	2	Cell radii within the Fresnel zone break point

We also note that all propagation models exhibit free-space path loss within the Fresnel zone break point. We then classified the propagation environments into four categories: dense urban, urban, suburban, and rural.

Each propagation environment has a unique path-loss slope. Table 2.1 provides a guideline to use these propagation models in various propagation environments.

2.4 A PC-Based RF Planning Tool

2.4.1 Background

Today, numerous computer-aided RF design tools are available for planning and designing the cellular system. These tools generally begin with empirical propagation models such as Okumura-Hata and Walfisch-Ikegami models, where the geographic information is already built-in. Yet, most prediction tools still require GIS tool to run the prediction. In addition, drive test data is used for model tuning, thus defeating the original purpose. Moreover, these tools require user-defined clutter factors, which are subjective; as a result an error is inevitably present in these tools. Last but not least, these tools are complex and expensive; users require specialized training to use them effectively.

The RF planning tool presented here differs from the others in that it is very inexpensive and simple to develop as a student project [8]. It is PC based and uses empirical propagation models, MS Excel, a GIS (Geographic Information Services) software, and a GPS (Global Positioning System) receiver. The GPS receiver will be used to collect the coordinates of each cell site (lat/long). Once the lat/long of antenna locations are known, the cell radii can be computed in Excel by using any of the existing empirical models such as Okumura-Hata or Walfisch-Ikegami. Finally, the MapInfo software will be used to import lat/long and the corresponding radius to display the RF coverage plot along with the geographic information such as roads, highways, water, park, buildings, etc. We believe that this project will be a valuable learning experience for the students. They will gain hands-on experience in project management, time management, teamwork, and documentation.

The RF planning tool presented below (Fig. 2.7) has been developed as a student project. It uses:

- A GIS software such as MapInfo
- Empirical propagation models
- MS Excel and
- PC

The GIS (Geographic Information Software) tool imports the cell radius. Empirical propagation models, such as Okumura-Hata, Walfisch-Ikegami, Lee, etc., provide path-loss characteristics in a given propagation environment. The MS Excel computes cell radius from a given propagation model. It also tabulates long/lat, antenna height, cell radii, etc. Upon receiving the lat/long, the PC-based RF prediction tool displays the RF coverage along with the geographic information.

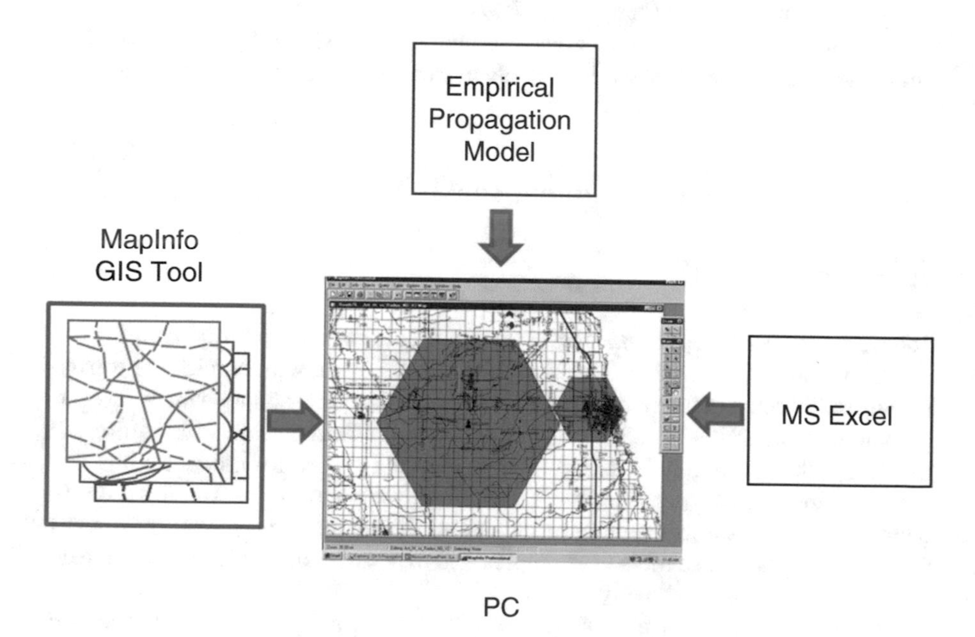

Fig. 2.7 Illustration of the PC-based RF planning tool

2.4.2 Step-By-Step Design Guide

Step 1: Collect latitude and longitude (lat/long) using GPS receiver, and determine base station antenna height. This is an outdoor activity involving:

- Site visits
- Identifying base station location
- Recording lat/long using GPS receiver
- Record antenna height

Step 2: Choose a propagation model from Table 2.1, and compute the required path loss L_p and the corresponding received signal:

- Path loss: $L_p = 134.2$ dB (from the chosen model)
- Received signal = EIRP – L_p (will vary from cell site to cell site)

Step 3: Tabulate design parameters. This step involves collecting and tabulating various design parameters listed below:

- Frequency
- Propagation model
- Base station power
- Antenna gain
- Signal strength at the cell edge
- Mobile antenna height
- Path loss L_p (from the propagation model)

Step 4: Compute cell radius as a function of antenna height. As an example, consider the Hata path-loss model:

- $L_p(\text{dB}) = C_1 + C_2 \log(f) - 13.82 \log(h_b) - a(h_m) + [44.9 - 6.55 \log(h_b)] \log(d) + C_o$
- Solving for cell radius (d), we obtain:
- Cell Radius: $d = 10^{\frac{L_p(\text{dB}) - C_o - [C_1 + C_2 \log(f) - 13.82 \log(h_b) - a(h_m)]}{[44.9 - 6.55 \log(h_b)]}}$

Step 5: Repeat Step 4 for each location (lat/long) and generate an Excel spreadsheet. This spreadsheet produces cell radii as a function of antenna height for each location. The table below shows the result for two locations.

Longitude	Latitude	Antenna height (m)	Cell radius (km)	Propagation environment
–97.0958	47.9202	15	6.069	Urban
–97.376	47.8978	30	12.961	Suburban

Step 6: Import lat/long and the corresponding cell radius from the above table into the GIS tool (e.g., MapInfo) using the specific set of commands, supported by the GIS software. It is assumed that the GIS software has already been installed into the

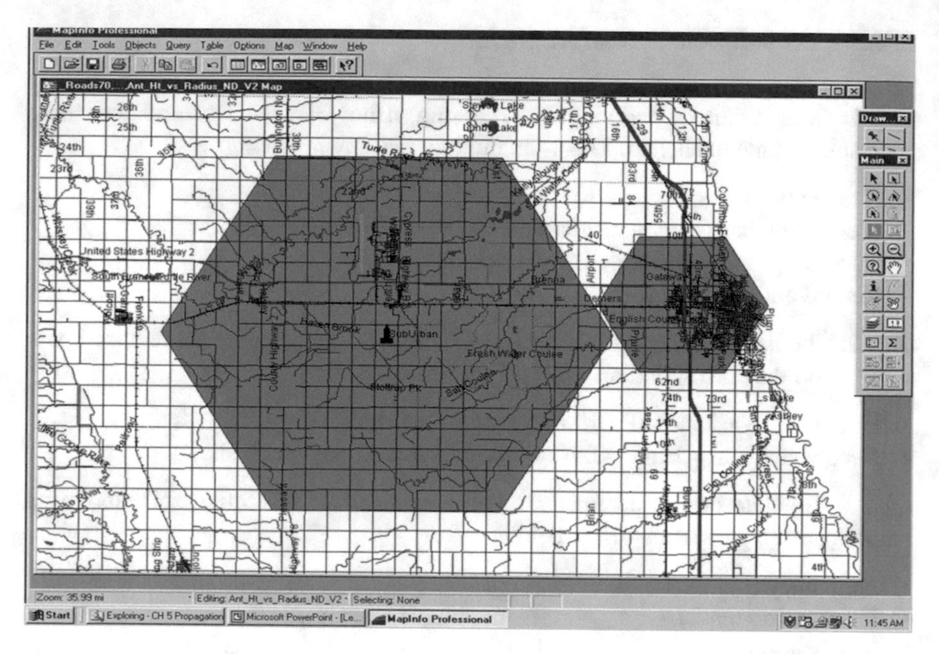

Fig. 2.8 Composite RF coverage plot on PC

PC and it is up and running. At this point, each lat/long and the corresponding cell radius will be imported instantly. The outcome is a composite RF coverage plot superimposed on the map, as shown in Fig. 2.8.

2.5 Conclusions

We have presented an overview of various empirical prediction models and have shown that these propagation models also exhibit equation of straight line within the Fresnel zone break point. Although these predictions and measurement techniques are the foundation of today's cellular services, they suffer from inaccuracies due to user-defined clutter factors. These clutter factors arise due to numerous RF barriers which vary from place to place. It is practically impossible to accommodate all these factors accurately. Cell site location is also a challenging engineering task because of regulations and restrictions imposed on some locations. Therefore, cell sites have to be relocated from the predicted location, requiring best judgment of RF engineers. Thus, we came to the conclusion that propagation prediction is a combination of science, engineering, and art. An experienced RF engineer, willing to compromise between theory and practice, is expected to accomplish the most. A simple and cost-effective propagation tool is presented as a student project.

References

1. S. Faruque, *Cellular Mobile Systems Engineering* (Artech House Inc., Norwood, MA, 1996). ISBN: 0-89006-518-7
2. M. Hata, Empirical formula for propagation loss in land mobile radio services. IEEE Trans. Veh. Technol. **VT-29**, 317–326 (1980)
3. J. Walfisch et al., A theoretical model of UHF propagation in urban environments
4. W.C.Y. Lee, *Mobile Cellular Telecommunications Systems* (McGraw-Hill Book Company, New York, 1989)
5. I. Miller, J.E. Freund, *Probability and Statistics for Engineers* (Prentice-Hall Inc., Englewood Cliffs, NJ, 1977)
6. D.A. Freedman, *Statistical Models: Theory and Practice* (Cambridge University Press, New York, 2005)
7. G.U. Yule, On the theory of correlation. J. R. Stat. Soc. (Blackwell Publishing) **60**(4), 812–854 (1897). https://doi.org/10.2307/2979746. JSTOR 2979746
8. S. Faruque, *Radio Frequency Propagation Made Easy*, Springer Briefs in Electrical and Computer Engineering, 1st edn. (Springer International Publishing, Cham, 2015). ISBN-13: 978-3319113937, ISBN-10: 3319113933

Chapter 3
Choice of Multiple Access Technique

Abstract Multiple access techniques enable many users to share the same spectrum in the frequency domain, time domain, code domain, or phase domain. These techniques are readily available as international standards and can be used to provide cellular services in a given geographical area. This chapter presents a brief overview of the various multiple access techniques available today. Numerous illustrations are used to bring students up-to-date in key concepts, underlying principles and practical applications of FDMA, TDMA, CDMA, and OFDMA. The mode of operation such as fDD and TDD is also presented with illustrations. To illustrate the concept, we present the construction of these radios.

3.1 Introduction to Multiple Access Techniques

The multiple access technique is well known in cellular communications [1–6]. It enables many users to share the same spectrum in the frequency domain, time domain, code domain, or phase domain. It begins with a frequency band, allocated by the FCC (Federal Communication Commission) [7]. The FCC provides licenses to operate wireless communication systems over given bands of frequencies. These bands of frequencies are finite and have to be further divided into smaller bands (channels) and reused to provide services to other users. This is governed by the International Telecommunication Union (ITU) [8]. ITU generates standards such as FDMA, TDMA, CDMA, OFDMA, etc., for wireless communications. Figure 3.1 illustrates the basic concept of various multiple access techniques currently in use.

In any multiple access technique, multiple users have access to the same spectrum, so that the occupied bandwidth does not exceed the FCC-allocated channel. Furthermore, as the size and speed of digital data networks continue to expand, bandwidth efficiency becomes increasingly important. This is especially true for broadband communication, where the choice of modulation scheme is important keeping in mind the available bandwidth resources, allocated by FCC.

With these constraints in mind, this chapter will present a comprehensive yet concise overview of multiple access techniques used in the cellular industry.

S. Faruque, *Radio Frequency Cell Site Engineering Made Easy*, SpringerBriefs in Electrical and Computer Engineering, https://doi.org/10.1007/978-3-319-99615-8_3

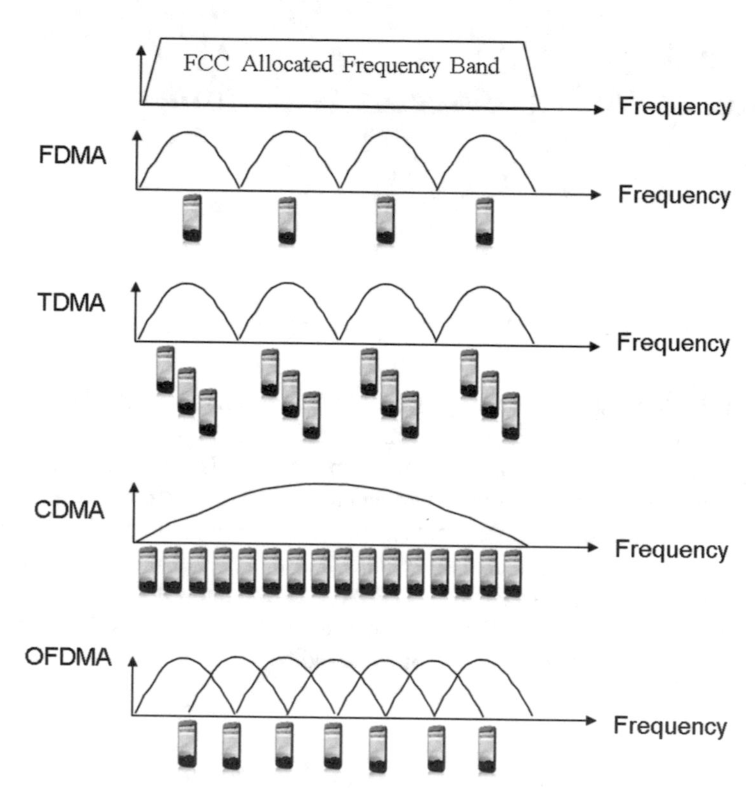

Fig. 3.1 Basic concept of multiple access techniques

3.2 Frequency Division Multiple Access (FDMA)

3.2.1 FDMA Concept

FDMA (Frequency Division Multiple Access) is the oldest communication technique, used in broadcasting, land-mobile two-way radio, etc. [1]. It begins with a band of frequencies, allocated by the FCC (Federal Communications Commission). The FCC provides licenses to operate wireless communication systems over given bands of frequencies. These bands of frequencies are further divided into several channels and assigned to users for full-duplex communication. Figure 3.2 illustrates the basic principle of a typical FDMA technique.

As shown in the figure, the FCC-allocated frequency band is divided into several frequencies, also known as channels. Each channel is assigned to a single user. In this scheme, the channel is occupied for the entire duration of the call. The communication link is maintained in both directions, either in the frequency domain or in the time domain. This is governed by two basic mode of operations listed below:

- Frequency division duplex (FDD)
- Time division duplex (TDD)

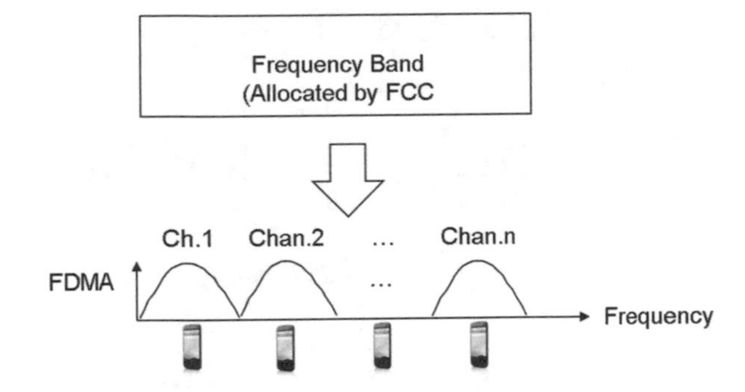

Fig. 3.2 Illustration of a typical FDMA technique. The FCC-allocated frequency band is divided into several channels. Each channel is assigned to a user

In FDD, all the available channels are divided into two bands, lower band and upper band, and grouped as pairs, one for the upload and the other for the download, separated by a guard band. As a result, both transmissions can take place at the same time without interference. This scheme is known as FDMA-FDD technique.

In TDD, a single frequency is time-shared between the uplink and the downlink. In this scheme, when the mobile transmits, base station listens, and when the base station transmits, the mobile listens. This is accomplished by formatting the data into a "frame," where the frame is a collection of several time slots. Each time slot is a package of data, representing digitized voice, digitized text, digitized video, and synchronization bits (sync bits). The sync bits are unique, which is used for frame synchronization. This scheme is known as FDMA-FDD technique.

3.2.2 FDMA-FDD Technique

In FDMA-FDD, all the available channels are divided into two bands, lower band and upper band, and grouped as pairs—L1U1, L2U2, …, LnUn. This is shown in Fig. 3.3. As can be seen in the figure, FDD uses two different frequencies, one for the upload and the other for the download, separated by a guard band. As a result, both transmissions can take place at the same time without interference. This scheme is known as FDMA-FDD technique.

A brief description of FDMA-FDD communication, as implemented in 1G cellular system [1–3], is presented below:

- The base station modulates the carrier frequency (U1) from the upper band and sends the modulated carrier to the mobile. The input modulating signal can be either analog or digital.
- Since the mobile is tuned to the same carrier frequency, it receives the modulated carrier from the base after a propagation delay. It then demodulates the carrier and recovers the information signal.
- In response, the mobile modulates a different carrier frequency (L1) from the lower band and transmits back to the base.
- The base station receives the modulated signal from the mobile and demodulates and recovers the information.
- The process continues until one of the transmitter terminates the call.

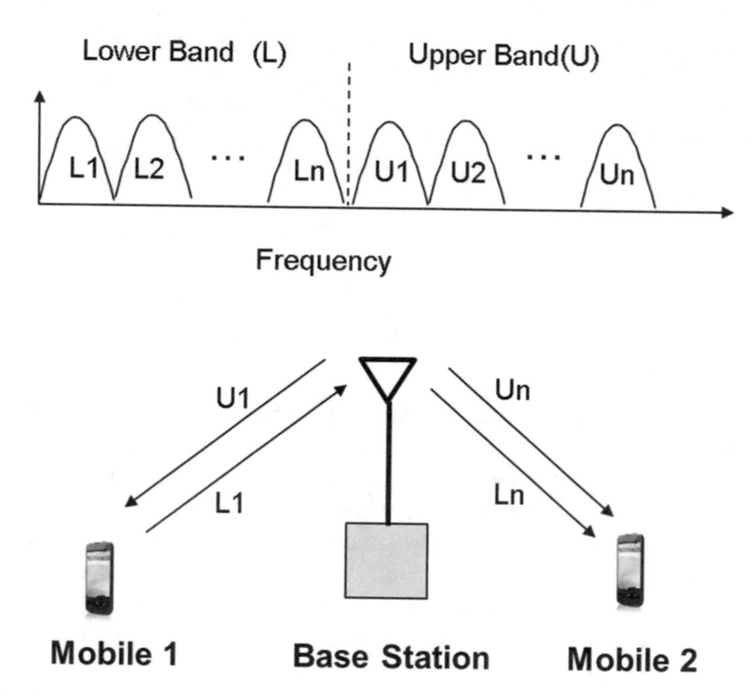

Fig. 3.3 Frequency division duplex (FDD) technique. Both frequencies can operate at the same time without interference

3.2.3 FDMA-TDD Technique

In FDMA-TDD, a single FDMA frequency is time-shared between the uplink and the downlink. The duration of transmission in each direction is generally short, in the order of ms. In this scheme, when the mobile transmits, base station listens, and when the base station transmits, the mobile listens. This is accomplished by formatting the data into a "frame," where the frame is a collection of several time slots. Each time slot is a package of data, representing digitized voice, digitized text, digitized video, and synchronization bits (sync bits). The sync bits are unique, which is used for frame synchronization. Figure 3.4 illustrates a typical frame and the TDD transmission scheme.

According to TDD transmission, both the base station and the mobile use the same carrier frequency. The transmit/receive mechanism between the base station and mobile is as follows:

- The base station modulates the carrier frequency by means of the digital information bits in frame F(B) and transmits to the mobile.
- Since the mobile is tuned to the same carrier frequency, it receives the frame F(B) after a propagation delay t_p.
- The mobile synchronizes the frame using the sync bits and downloads the data.

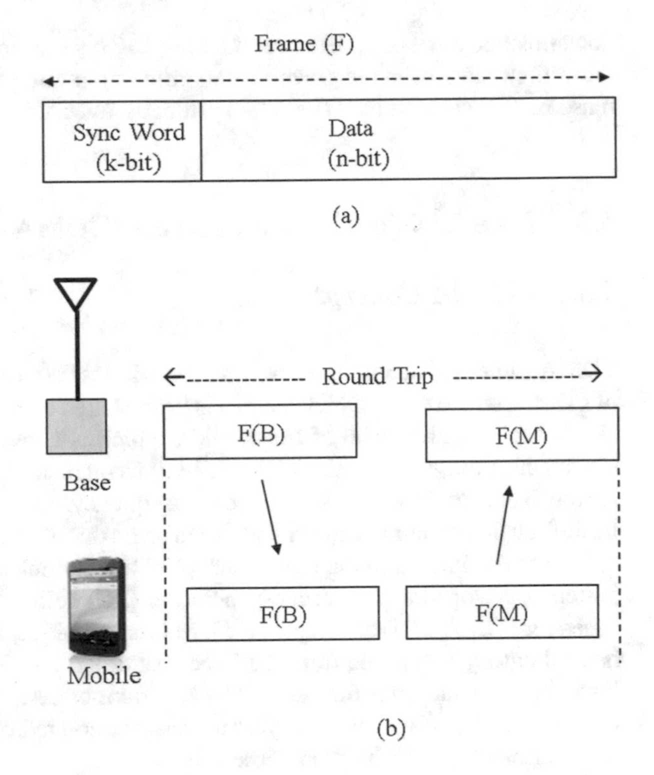

Fig. 3.4 TDD frame structure. (**a**) Frame. (**b**) Frame transmission scheme

- After a guard time t_g, mobile transmits its own frame F(M) to the base using the same carrier frequency.
- Base receives the frame from the mobile after a propagation delay t_p, maintains sync using the sync bits, and downloads the data.
- The round trip communication is now complete.
- The communication continues until one terminates the call.

As can be seen in the figure, the TDD schemes require a propagation delay and a guard time between transmission and reception. The complete round trip delay T_d must be sufficient to accommodate the frame, propagation delay, and the guard time. Therefore, the round trip delay can be written as:

$$T_d = 2\left(F + t_p + t_g\right) \tag{3.1}$$

The round trip delay T_d depends on the frame length F, which is generally in milliseconds (ms). The propagation delay t_p depends on the propagation distance, and the guard time t_g depends on the technology.

In 4G cellular communications, such as OFDMA and LTE, the traffic in both directions is not balanced. The volume of data transmission can be dynamically adjusted in each direction by means of the TDD technique. There is scheduling protocol, which can be dynamically controlled to offer high-speed data over the downlink and low-speed data over the uplink. This is accomplished by transmitting more time slots over the downlink, thereby supporting more capacity. For these reasons, TDD is used in 4G cellular system as WiMAX and LTE standards [4, 5].

3.3 Time-Division Multiple Access (TDMA)

3.3.1 TDMA Concept

TDMA (time-division multiple access) for wireless communication is an extension of FDMA, where each FDMA channel is time-shared by multiple users, one at a time [2]. It begins with a band of frequencies, which is allocated by the FCC (Federal Communications Commission). This band of frequencies further divided into several narrow bands of frequencies, where each frequency, also known as channel, is used for full-duplex communication by multiple users one at a time as depicted in Fig. 3.5.

Figure 3.6 illustrates the basic concept of a full-duplex cellular communication system, developed as the second-generation (2G) cellular communication system, based on TDMA-FDD technique [1–5]. In this technique, a pair of FDMA channels is used during a call, one from the lower band and one from the upper band. The lower band frequency is time-shared by several mobiles. The upper band frequency is also time-shared synchronously by the base station radio. Both channels are occupied during the entire duration of the call.

Synchronization is achieved by means of a special frame structure, where the frame is a collection of time slots. Each time slot is assigned to a mobile. This implies that when one mobile has access to the channel, the other mobiles are idle. Therefore, TDMA synchronization is critical for data recovery and collision avoidance.

TDMA has several advantages over FDMA:

- Increased channel capacity
- Greater immunity to noise and interference
- Secure communication
- More flexibility and control

Moreover, it allows the existing FDMA standard to coexist in the same TDMA platform, sharing the same RF spectrum.

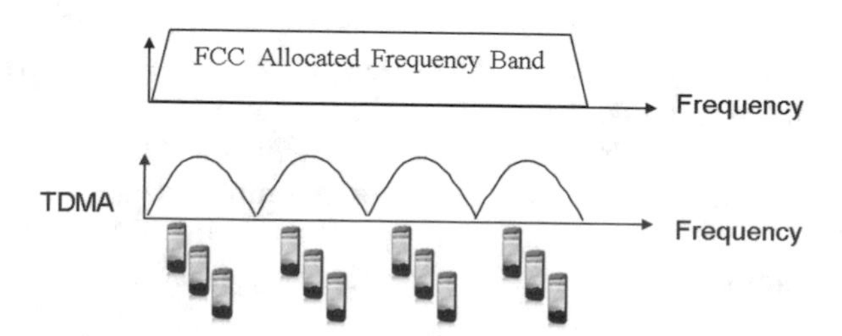

Fig. 3.5 North American cellular TDMA technique. The FCC-allocated frequency band is divided into several channels where each channel is time-shared by several users one at a time

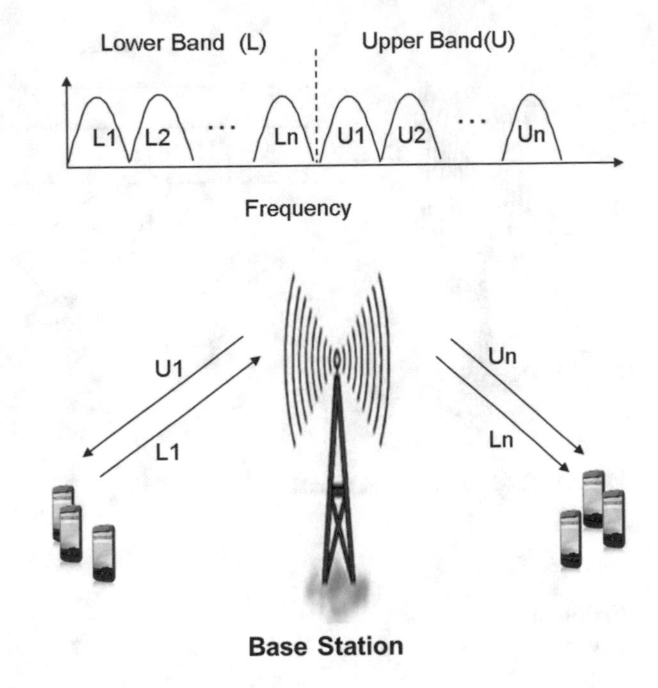

Fig. 3.6 North American 2G cellular TDMA communication system

3.3.2 TDMA Frame Structure

The North American 2G TDMA air link is based on a 40 ms frame structure, equally divided into six time slots, 6.667 ms each. Each of the six time slots contains 324 gross bit intervals, corresponding to 162 symbols (π/4 DQPSK modulation, 1 symbol = 2 bits of information). Figure 3.7 shows the forward link (base to mobile) TDMA frame structure. In TDMA-3, the time slots are paired as 1–4, 2–5, and 3–6 where each disjointed pair of time slots is assigned to a mobile. This arrangement enables three mobiles to access the same 30 kHz channel one at a time.

The TDMA-3 forward link uses a rate 1/2 convolutional encoding with interleaving. The encoded 48.6 kb/s data bit stream is modulated by means of a π/4 DQPSK modulation and then transmitted from the base station to the mobile where each mobile receives data at 16.2 kb/s. At the receive side, the RF signal is demodulated and decoded, and finally the original data is recovered. Since this is a radio channel, the recovered data is impaired by noise, interference, and fading. As a result, the information is subject to degradation. Although error control coding greatly enhances the performance, the C/I (carrier-to-interference ratio) is still the limiting factor. The TDMA-3 reverse link is exactly the reverse process.

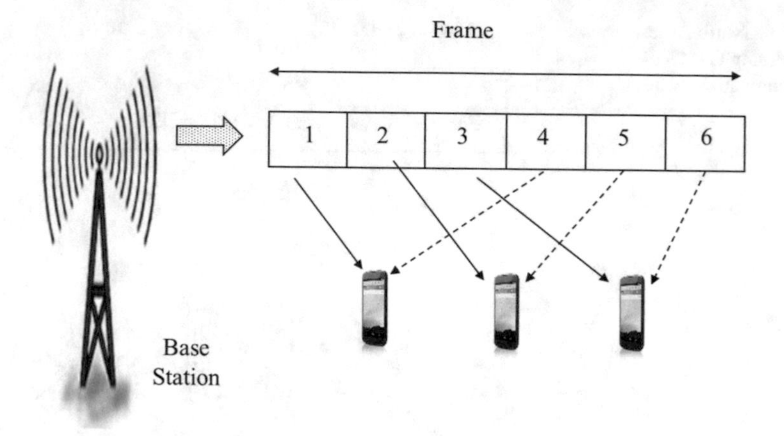

Fig. 3.7 TDMA forward link format

Problem

Given:

- Frame length = 40 ms (figure below).
- The frame contains six time slots and supports three users.
- Each user originates 16.2 kb/s data.

Frame =40 ms

1	2	3	4	5	6

Find:

(a) A suitable multiplexing structure
(b) The composite data rate in the channel
(c) Number of bits/frame

Solution:

(a) We have three users and six time slots. Therefore, we can assign two time slots/user:

- User 1: Time slots 1 and 4
- User 2: Time slots 2 and 5
- User 3: Time slots 3 and 6

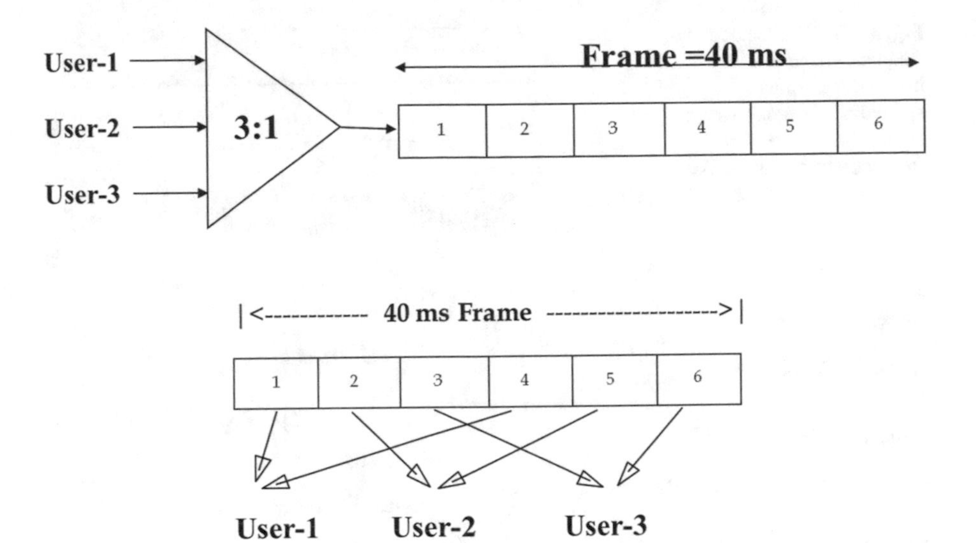

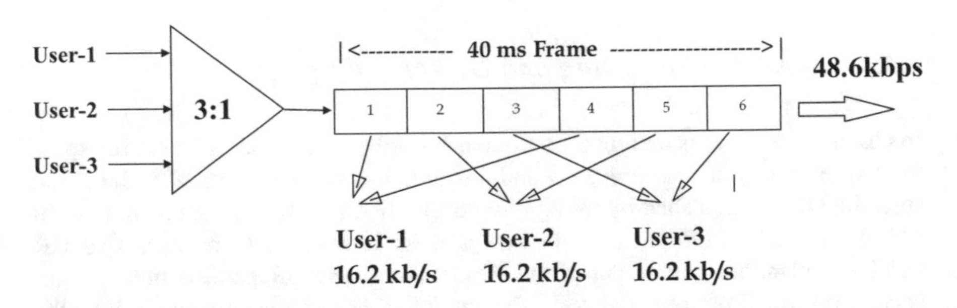

(b) Composite data rate: 16.2 kb/s/user × 3 = 48.6 kb/s

(c) Number of bits/frame = frame length/bit duration = 40 ms/(1/48.6 kb/s) = 1944 bits/frame

3.4 Code-Division Multiple Access (CDMA)

3.4.1 CDMA Concept

CDMA (code-division multiple access) is a spread spectrum (SS) communication system where multiple users have access to the same career frequency at the same time [3]. It begins with a frequency band, allocated by the Federal Communication Commission (FCC) as shown in Fig. 3.8. The FCC provides licenses to operate wireless communication systems over given bands of frequencies. These frequency bands are finite and have to be reused to support a large number of users in a given geographical area.

The objective of this chapter is to review spectrum, spectrum spreading, and de-spreading techniques and show how it relates to spread spectrum CDMA radio.

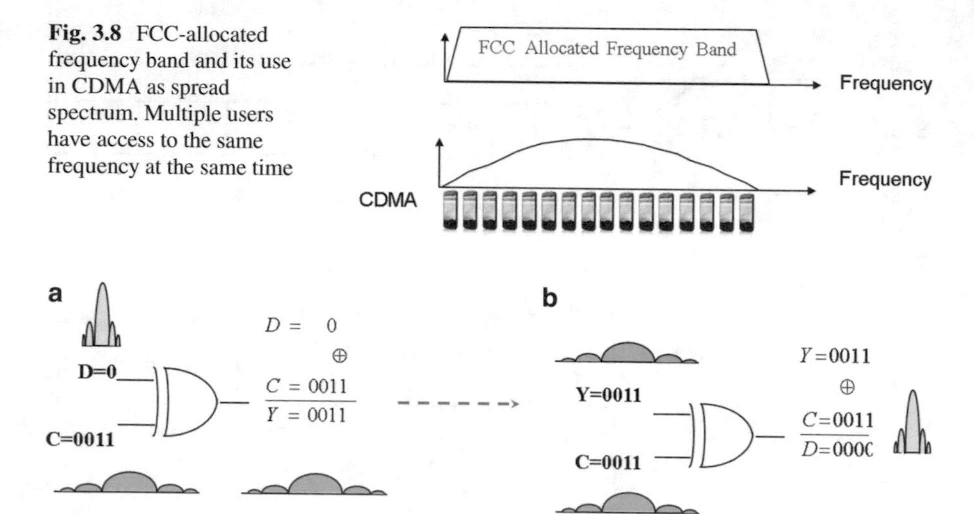

Fig. 3.8 FCC-allocated frequency band and its use in CDMA as spread spectrum. Multiple users have access to the same frequency at the same time

Fig. 3.9 Spreading and de-spreading technique. (**a**) Spreading bit 0. (**b**) De-spreading and recovering bit 0

3.4.2 Spectrum Spreading and De-spreading

In CDMA, each user is assigned a unique n-bit orthogonal code as a user ID, spectrum spreading at the transmit side, and de-spreading at the receive side. Spectrum spreading is accomplished by multiplying each NRZ data bit by means of an n-bit orthogonal code. Multiplication in this process is referred to as exclusive OR (EXOR) operation. The output of the EXOR (exclusive OR gate) is now a high-speed orthogonal or antipodal code. De-spreading is a similar process, where the receiver multiplies the incoming data by means of the same orthogonal code and recovers the data. Let us examine these operations using a 4-bit orthogonal code, where the code sequence is given by 0011 and the input NRZ data is 0 and 1.

Example 1: Spreading Bit 0 and De-spreading to Recover Bit 0 This is shown in Fig. 3.9. When the binary bit 0 is multiplied by a 4-bit orthogonal code 0011, we write:

$$0 \text{ EXOR } (0011) = 0011 \tag{3.2}$$

This is the orthogonal code, reproduced due to exclusive OR operation. Moreover, the bit rate is also multiplied by a factor of 4, thereby spreading the spectrum by a factor of 4 as well. This is the wide band data which is transmitted to the receiver. Upon receiving 0011, the receiver performs the de-spreading function using the same orthogonal code 0011, which is also an EXOR function. Thus we write:

$$0011 \text{ EXOR } 0011 = 0000$$

This is the original data bit 0, which has been reproduced due to exclusive OR operation. Moreover, the bit rate is also divided by a factor of 4.

Example 2: Spreading Bit 1 and De-spreading to Recover Bit 1 This is shown in Fig. 3.10. When the binary bit 1 is multiplied by the same orthogonal code, we obtain:

$$1\ \text{EXOR}\ (0011) = 1100 \tag{3.3}$$

This is the inverse of the orthogonal code. This code is also known as antipodal code. The spectrum is also spread by a factor of 4.

Upon receiving the antipodal code 1100, the receiver performs the de-spreading function using the same orthogonal code 0011, which is also an EXOR function. Thus we write 1100 EXOR 0011 = 1111. This represents the original data bit 1, which has been reproduced due to exclusive OR operation. Moreover, the bit rate is also divided by a factor of 4.

3.4.3 Construction of CDMA Radio

Let us examine a CDMA radio based on 4-bit orthogonal code, where the code sequence is given by 0101. This is shown in Figs. 3.11 and 3.12.

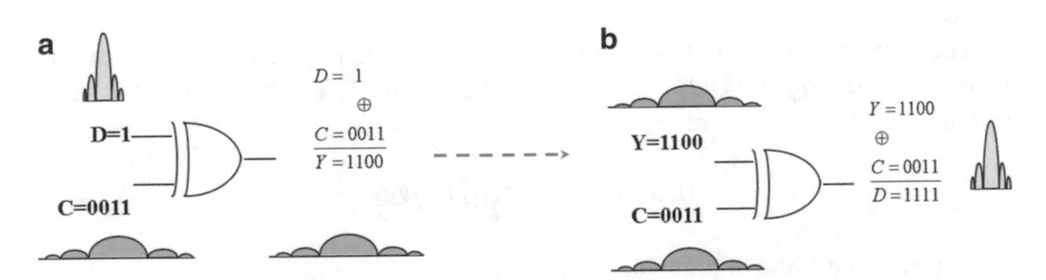

Fig. 3.10 Spreading and de-spreading technique. (**a**) Spreading bit 1. (**b**) De-spreading and recovering bit 1

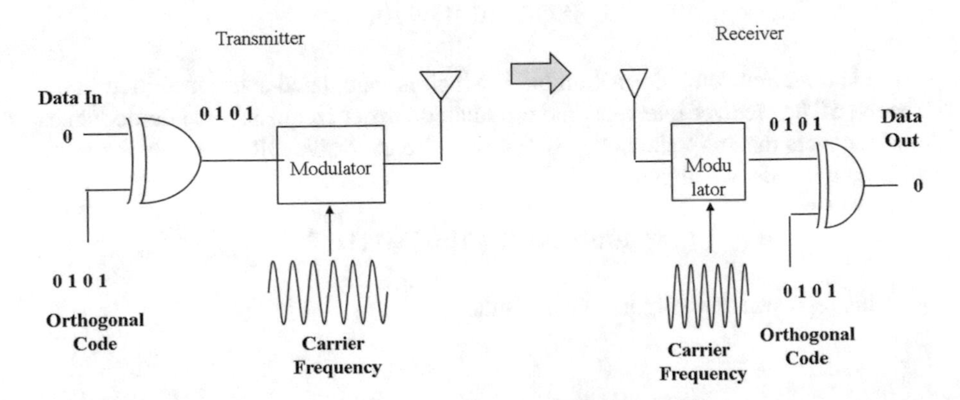

Fig. 3.11 CDMA radio. Binary bit 0 transmit/receive mechanism

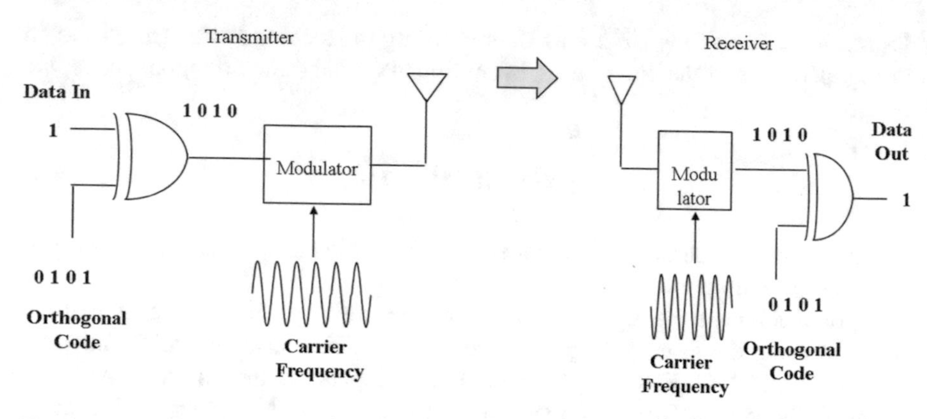

Fig. 3.12 CDMA radio. Binary bit 1 transmit/receive mechanism

In Fig. 3.11, the binary bit 0 is multiplied by the 4-bit orthogonal code to reproduce the orthogonal code as follows:

$$(\ )\text{EXOR }(0101) = 0101 \tag{3.4}$$

This represents the information bit 0, which is modulated and transmitted to the receiver.

The receiver intercepts the modulated carrier frequency and demodulates and recovers the orthogonal code 0101. Since the exclusive OR gate also uses the same orthogonal code, we obtain:

$$0101\ \text{EXOR }(0101) = 0000$$

This represents the original binary value 0.

Similarly, in Fig. 3.12, the binary bit 1 is multiplied by the same 4-bit orthogonal code to produce the antipodal code as follows:

$$1\ \text{EXOR }(0101) = 1010 \tag{3.6}$$

This represents the information bit 1, which is modulated and transmitted to the receiver. The receiver intercepts the modulated carrier frequency and demodulates and recovers the antipodal code 1010. Since the exclusive OR gate uses the same orthogonal code, we obtain:

$$1010\ \text{EXOR }(0101) = 1111$$

This represents the original binary value 1.

In summary, CDMA is a branch of multiple access techniques, where multiple users have access to the same spectrum through orthogonal codes. Orthogonal codes are binary valued and have equal number of 1s and 0s. Therefore, for an n-bit orthogonal code, there are n orthogonal codes. In CDMA, each user is assigned a unique orthogonal code. As a result, each user remains in orthogonal space after modulation and offers maximum isolation. Yet, there is a limit to the use of all the codes, which is related to channel capacity.

3.5 Orthogonal Frequency-Division Multiple Access (OFDMA)

3.5.1 OFDMA Concept

OFDMA (orthogonal frequency-division multiple access) [4, 5] is a relatively new wireless communication standard used in 4G-WiMAX (Worldwide Interoperability for Microwave Access) and 4G-LTE (long-term evolution) protocol. It may be noted that WiMAX is an IEEE 803.16 standard while LTE is a standard developed by the 3GPP group. Both standards are surprisingly similar and bandwidth efficient. OFDMA is used in the 4G cellular standard.

In OFDMA, each frequency is placed at the null of the adjacent frequency (see Fig. 3.13). This is governed by the well-known "Fourier transform," so that adjacent frequencies are orthogonal to each other. It begins with a band of frequencies. This band of frequencies is allocated by the FCC (Federal Communications Commission). This band of frequencies is further divided into several narrow bands of frequencies, where each frequency is orthogonal to each other. OFDMA is a full-duplex communication system. The communication link is maintained in both directions in the time domain known as time-division duplex (TDD).

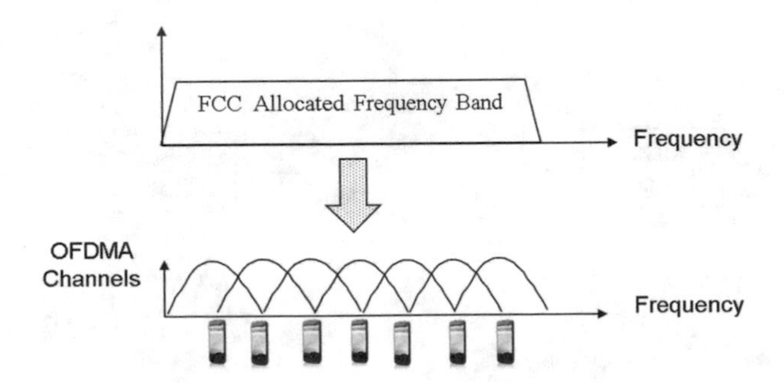

Fig. 3.13 Construction of OFDMA channels from FCC-allocated frequency band. There are n − 1 nulls in the FCC-allocated band, where n is the number of FDMA channels

3.5.2 *OFDMA Radio and Spectrum Allocation*

Figure 3.14 shows the basic OFDMA radio and spectrum allocation scheme. Here, each frequency band is placed at the null of the adjacent frequency band, where each frequency is orthogonal to each other. The adjacent bans are determined by the bit rate and modulation.

Since OFDMA is a full-duplex multiuser wireless communication system, a carrier frequency is assigned to a pair of users to operate in the TDD mode. In this scheme, the channel is occupied by two users for the entire duration of the call. The communication link is maintained in both directions in the time domain known as time-division duplex (TDD).

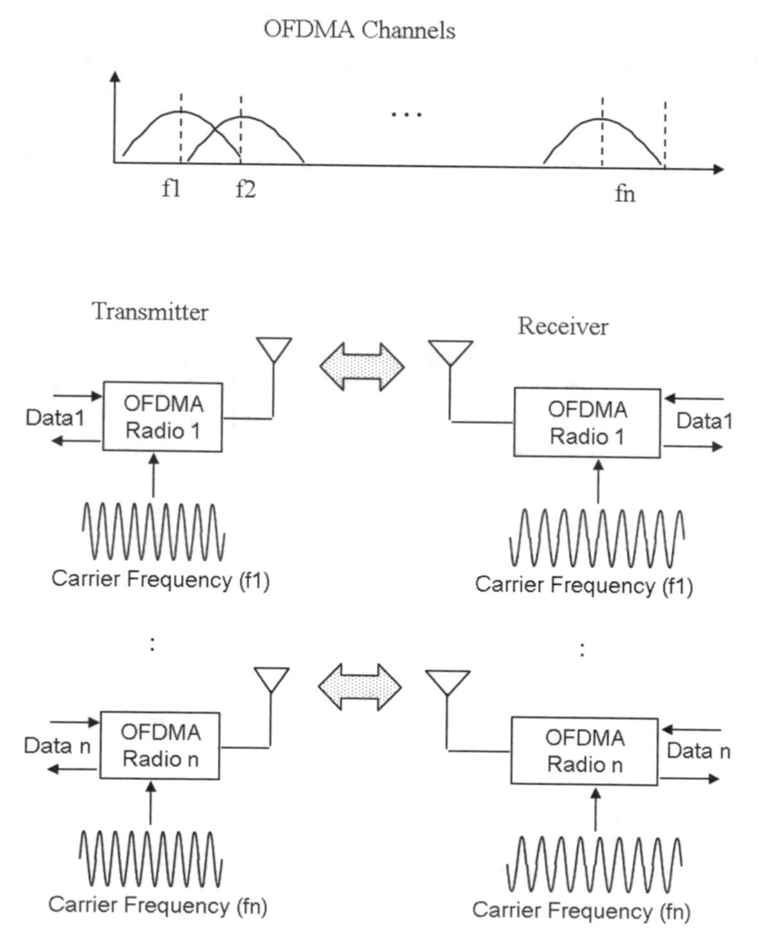

Fig. 3.14 OFDMA radio and spectrum allocation scheme. A single frequency is used in both directions. The mode of operation is TDD

3.5.3 OFDMA Channel Capacity

The bandwidth of each OFDMA channel is determined by the first null to first null of the two-sided response of the main lobe. Each frequency band is placed at the null of the adjacent band as shown in the figure. Notice that there are n − 1 nulls in the OFDMA spectrum, where n is the number of FDMA channels. Therefore, the number of OFDMA channels will be given by:

$$\text{Number of OFDMA Channels} = 2 \times \text{FDMA Channels} - 1 \tag{3.7}$$

3.5.4 OFDMA-TDD Mode of Operation

OFDMA is a full-duplex communication system. The communication link is maintained in both directions in the time domain known as time-division duplex (TDD). In OFDMA-TDD, a single frequency is time-shared between the uplink and the downlink. The duration of transmission in each direction is generally short, in the order of ms (millisecond). In this scheme, when the mobile transmits, base station listens, and when the base station transmits, mobile listens. This is accomplished by formatting the data into a "frame," where the frame is a collection of several time slots. Each time slot is a package of data, representing digitized voice, digitized text, digitized video, and synchronization bits (sync bits). These unique sync bits are used for synchronization.

Figure 3.15 illustrates a typical frame and the TDD transmission scheme. According to TDD transmission, both the base station and the mobile use the same carrier frequency. The transmit/receive mechanism between the base station and mobile is as follows:

- The base station modulates the carrier frequency by means of the digital information bits in frame F(B) and transmits to the mobile.
- Since the mobile is tuned to the same carrier frequency, it receives the frame F(B) after a propagation delay t_p.
- The mobile synchronizes the frame using the sync bits and downloads the data.
- After a guard time t_g, the mobile transmits its own frame F(M) to the base using the same carrier frequency.
- Base receives the frame from the mobile after a propagation delay t_p, maintains sync using the sync bits, and downloads the data.
- A round trip communication is now complete.
- The communication continues until one terminates the call.

As can be seen in the figure, the TDD schemes require a propagation delay and a guard time between transmission and reception. The complete round trip delay T_d must be sufficient to accommodate the frame, propagation delay, and the guard time. Therefore, the round trip delay can be written as:

$$T_d = 2\left(F + t_p + t_g\right) \tag{3.8}$$

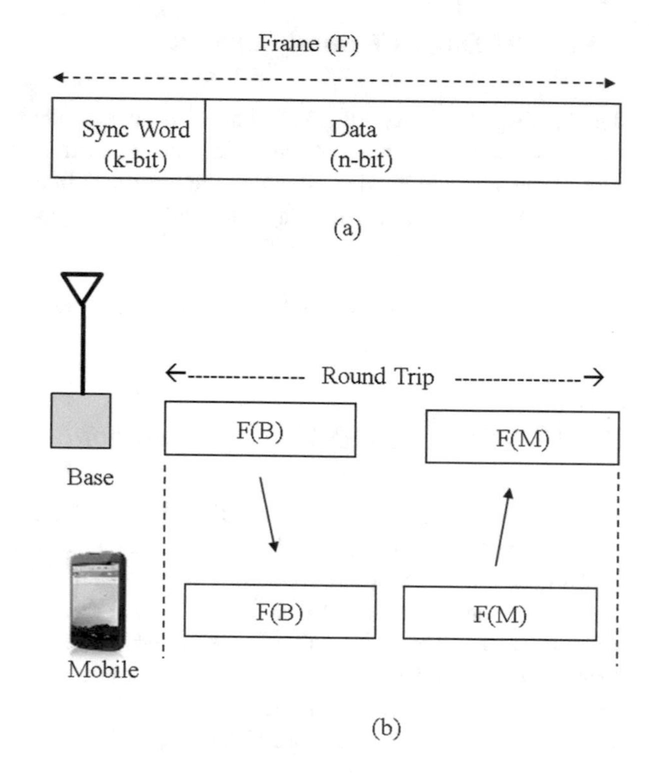

Fig. 3.15 OFDMA-TDD frame structure. (**a**) Frame. (**b**) Frame transmission scheme

The round trip delay T_d depends on the frame length F, which is generally in milliseconds (ms). The propagation delay t_p depends on the propagation distance, and the guard time t_g depends on the technology.

Problem

Given:

- Frame length = 2 ms
- Guard Time $t_g = 0.01$ ms
- Distance between the base station and mobile = 1 km
- Velocity of light $c = 3 \times 10^8$ m/s

Find:

- The round trip delay t_p.

Solution:

- Propagation delay for a distance of 1 km = 1 km/(3×10^8 m) = 3.3 ms
- Round trip delay $t_d = 2(t_p + t_f + t_g) = 2(3.3 \text{ ms} + 2 \text{ ms} + 0.01 \text{ ms}) = 5.31$ ms

3.6 Conclusions

Multiple access techniques enable many users to share the same spectrum in the frequency domain, time domain, code domain, or phase domain. It begins with a frequency band, allocated by FCC (Federal Communication Commission). FCC provides licenses to operate wireless communication systems over given bands of frequencies. These bands of frequencies are finite and have to be further divided into smaller bands (channels) and reused to provide services to other users. This is governed by the International Telecommunication Union (ITU). ITU generates standards such as FDMA, TDMA, CDMA, OFDMA, etc., for wireless communications. These standards are readily available and can be used to provide cellular services in a given geographical area.

This chapter presents a brief overview of various multiple access techniques used for cellular communications. Numerous illustrations are used to bring students up-to-date in key concepts, underlying principles, and practical applications of FDMA, TDMA, CDMA, and OFDMA. The mode of operation such as fDD and TDD is also presented with illustrations. Construction of these radios is also described for basic understanding.

References

1. Advanced Mobile Phone Services, Special Issue, Bell System Technical Journal, vol. 58, Jan 1979
2. IS-54, Dual-Mode Mobile Station-Base Station Compatibility Standard EIA/TIA Project Number 2215, Dec 1989
3. IS-95, Mobile Station—Base Station Compatibility Standard for Dual Mode Wide Band Spread Spectrum Cellular Systems, TR 45, PN-3115, 15 Mar 1993
4. ITU-R, Report M.2134, Requirements related to technical performance for IMT-Advanced radio interface(s), Approved in Nov 2008
5. S. Parkvall, D. Astely, The evolution of LTE toward LTE advanced. J. Commun. **4**(3), 146–154 (2009). https://doi.org/10.4304/jcm.4.3.146-154
6. S. Faruque, *Cellular Mobile Systems Engineering* (Artech House Inc., Norwood, MA, 1996). ISBN: 0-89006-518-7
7. FCC, Federal Communication Commission, Washington, DC, USA
8. ITU, International Telecommunications Union, Paris, France

Chapter 4
Frequency Planning

Abstract In frequency planning, we dole out the channels to the cells, much like a dealer in card game deals out cards from the deck until every player has a set. Beginning with the FCC-allocated frequency band, we show how channels are created and allocated to different cells. Cell reuse plans are described next with C/I. The improvement of channel capacity by means of optimization is also described.

4.1 Introduction

The FCC (Federal Communications Commission) provides licenses to operate cellular communication systems over given bands of frequencies [1–5]. These bands of frequencies are further divided into several smaller frequency channels and assigned to users by means of a technique known as frequency planning. In frequency planning, we dole out the channels to the cells, much like a dealer in card game deals out cards from the deck until every player has a set. Figure 4.1 shows the basic concept of frequency planning for a cluster of three cells.

Since the number of channels is finite, they have to be reused to provide services to other geographic areas [4, 6]. For example, in a cluster of three cells (as shown in Fig. 4.1), all the available channels (nine channels) are distributed among a cluster of three cells, three channels per cell. Next, each cluster of cells, along with the allocated channels, is placed adjacent to each other to serve adjacent geographical areas. This process is repeated to serve a given geographical area as required. This forms the basis of frequency planning.

This chapter provides a comprehensive yet a concise overview of the fixed spectrum management technique, used in the cellular industry. Beginning with the concept of the cell, this chapter briefly describes the cell geometry, followed by the concept of cell reuse with the evaluation of carrier-to-interference ratio (C/I). The classical cell reuse plan [3] is described next, with examples of various frequency plans related to OMNI and sectorization schemes. The improvement of channel capacity by means of optimization is also briefly described.

S. Faruque, *Radio Frequency Cell Site Engineering Made Easy*, SpringerBriefs in Electrical and Computer Engineering, https://doi.org/10.1007/978-3-319-99615-8_4

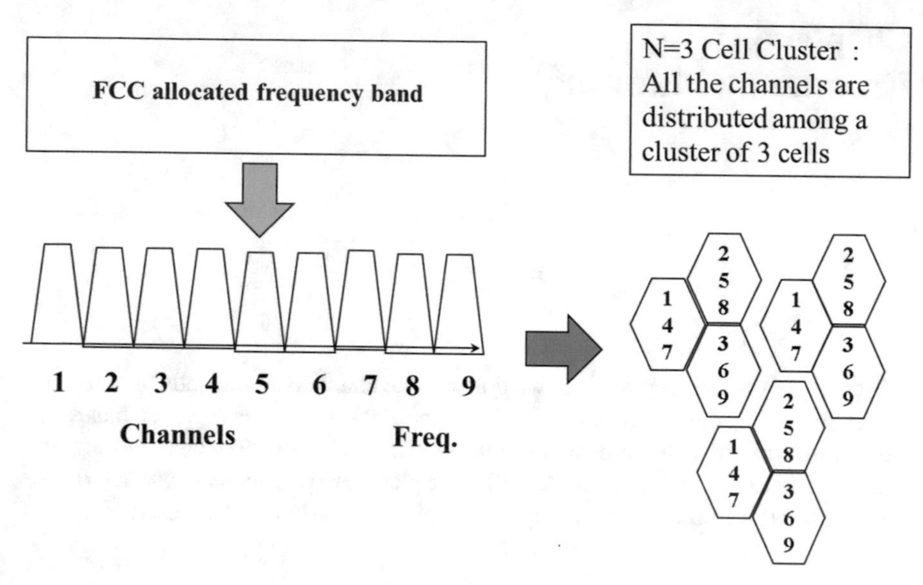

Fig. 4.1 Illustration of frequency planning. In frequency planning, we dole out channels among a cluster of a cell, much like a card game

4.2 Frequency Planning Techniques

4.2.1 Background

Several frequency planning or channel assignment techniques are available. Some of the most widely used frequency planning techniques are given below [5]:

- N = 7 frequency reuse plan
- N = 3 frequency reuse plan

The N = 7 is the classical cellular architecture, having seven hexagonal cells per cluster. All the available channels are evenly distributed among seven cells. This is the first arrangement which works in most propagation environments, giving 18+ dB C/I. It was originally developed by V.H. MacDonald in 1979 [4]. It ensures adequate channel reuse distance to an extent where co-channel interference is low and acceptable while maintaining a high channel capacity.

Similarly, N = 3 is the contemporary cellular architecture having three hexagonal cells per cluster. All the available channels are evenly distributed among three cells. Therefore, N = 3 frequency plan offers more capacity per cell.

These bands of frequencies are further divided into several channel groups and assigned to a cluster of cells, as shown in Fig. 4.1 for frequency reuse in clusters of 3s, known as N = 3 frequency reuse plan. This is the most widely used frequency reuse plan today. There are also other frequency reuse plans, for example, N = 4, N = 7, N = 9, etc. However, an increase in the number of cells per cluster decreases cell capacity.

4.2.2 N = 7 Frequency Planning

The N = 7 is the classical cellular architecture, which is based on hexagonal geometry. It was originally developed by V.H. MacDonald in 1979. It ensures adequate channel reuse distance to an extent where co-channel interference is low and acceptable while maintaining a high channel capacity. These frequency plans are briefly presented to illustrate the concept.

The scheme is shown in Fig. 4.2, where we have a cluster of seven OMNI cells and a cluster of seven sectorized cells. OMNI cell uses OMNI directional (all directions) antennas. In the sectorized scheme, each cell is divided into three sectors, 120° each. Directional antennas are used in each sector. According to the art of channel assignment technique, all the available channels are grouped into 21 frequency groups as follows:

G1, G2, G3, G4, G5, G6, G7, G8, G9, G10, G11, G12, G13, G14, G15, G16, G17, G18, G19, G20, and G21.

Each frequency group has several frequencies (known as channels). These 21 frequency groups are equally distributed among the cells/sectors. Notice that each OMNI cell gets 3 frequency groups or a total of 21 frequency groups in the 7-cell cluster. On the other hand, a sectorized cell gets one frequency group per sector for a total of three frequency groups per cell. The total number of frequency groups per cluster is still the same as in the OMNI scheme.

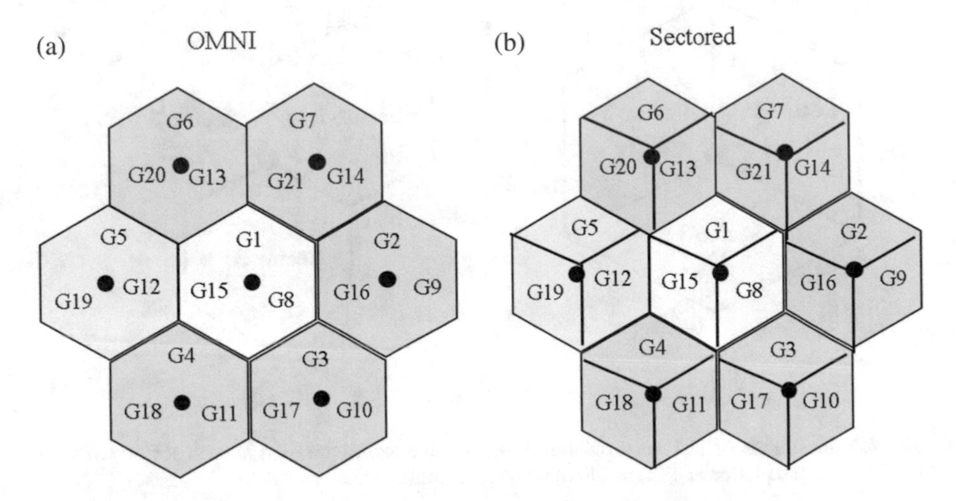

Fig. 4.2 The classical seven-cell cluster. (**a**) OMNI pattern and (**b**) sectored pattern

4.2.3 N = 7 Sectorization

The 120° sectorization is achieved by dividing a cell into three sectors, 120° each, as shown in Fig. 4.3a. Each sector is treated as a logical OMNI cell, where directional antennas are used in each sector for a total of three antennas per cell. Figure 4.3b shows an alternate representation, which is known as tri-cellular plan [5, 6]. Both configurations are conceptually identical, while the latter is convenient for channel assignment. Each sector uses one control channel and a set of different voice channels. Adequate channel isolations are maintained within and between sectors in order to minimize interference. This is attributed to channel assignment techniques, as we shall see later in this chapter.

Because directional antennas are used in sectored cells, it allows reuse of channels more frequently, thus enhancing channel capacity. Moreover, multipath components are also reduced due to antenna directivity, hence enhancing the C/I performance.

The growth plan is based on the distribution of 1 frequency group per sector, 3 frequency groups per cell for a total of 21 frequency groups per cluster. This is shown in Fig. 4.4a for the conventional plan and in Fig. 4.4b for the tri-cellular plan where the sector is represented by hexagon. Channel distribution is based on N, N + 7, N + 14 scheme where N = 1, 2, …, 7. Therefore for N = 1, cell 1 uses frequency group 1 for sector 1, frequency group 8 for sector 2, and frequency group 15 for sector 3. Similarly, cell 2 uses frequency group 2, 9, and 16 for sector 1, 2, and 3, respectively.

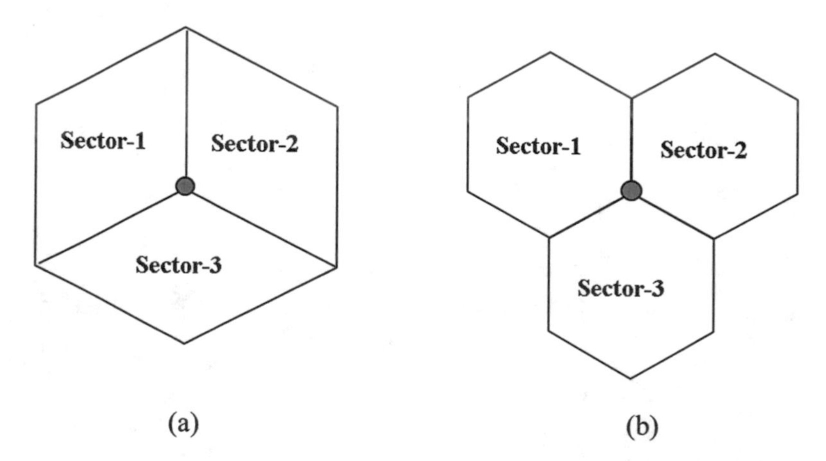

Fig. 4.3 Illustration of 120° sectorization. Directional antennas are used in each sector: (**a**) conventional representation and (**b**) an alternate representation

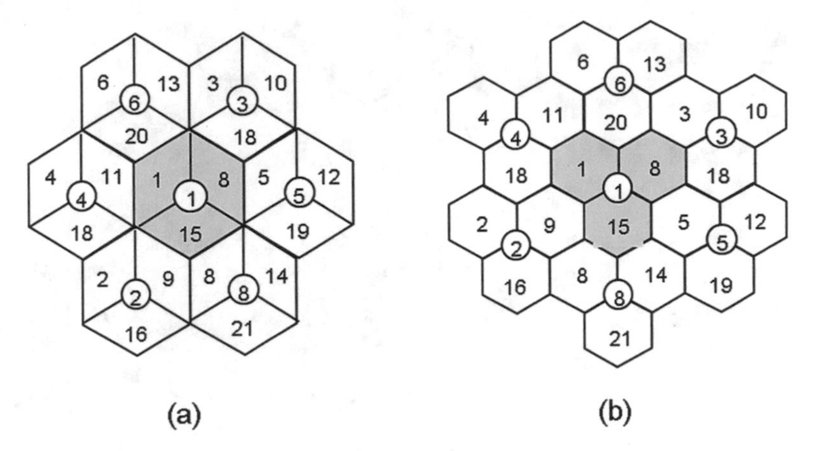

Fig. 4.4 (**a**) N = 7/21 sectorized configurations based on 120° sectorized plan and (**b**) N = 7/21 tri-cellular plan

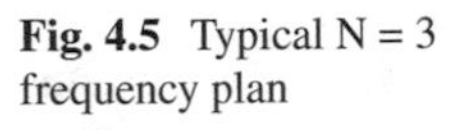

Fig. 4.5 Typical N = 3 frequency plan

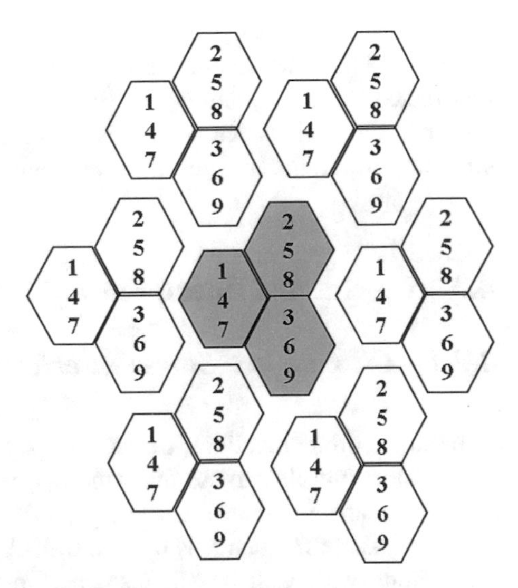

4.2.4 *N = 3 Tri-cellular Plan*

The N = 3 frequency plan is based on a cluster of three cells as shown in Fig. 4.5.

In this scheme, the FCC-allocated frequency band is divided into nine frequency channels. These frequency channels are then distributed among three cells much like a dealer in a card game deals out cards from the deck until every player has a set. In this case, each cell receives three channels (one channel/sector), according to this distribution plan. The growth plan is based on the allocation of the same frequency group among adjacent clusters (see Fig. 4.5).

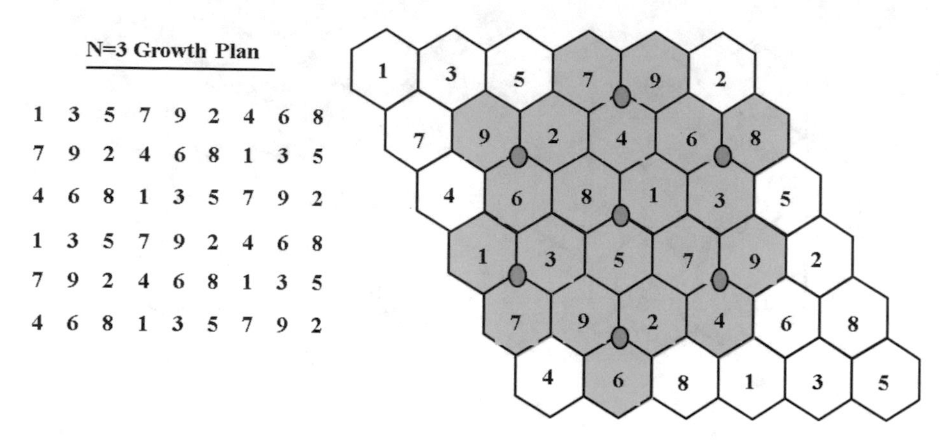

Fig. 4.6 N = 3, tri-cellular growth plan showing N = 7 mapping

Figure 4.6 shows another scheme where a 3 × 3 array of nine frequency groups distributed alternately among nine logical cells as shown in Fig. 4.6 [5]. Because of alternate channel assignment, this arrangement completely eliminates adjacent channels from adjacent sites, thus reducing adjacent channel interference.

4.3 Co-channel Interference (C/I)

4.3.1 C/I Due to a Single Interferer

In cellular communications, frequencies are reused in different cells, which means that another mobile can use the same frequency, thereby causing co-channel interference or carrier to interference (C/I) [4–6]. As an illustration, we consider Fig. 4.7, where the same frequency is used in Cell-A and Cell-B. Therefore, a mobile communicating with Cell-A will also receive the same frequency from the distant Cell-B. This is analogous to the "near-far" problem, causing co-channel interference. We use the following method to determine this interference. Let:

RSL_A = Received signal level at the mobile from Cell-A
d_A = Distance between the mobile and Cell-A
RSL_B = The received signal level at the mobile from Cell-B
d_B = Distance between the mobile and Cell-B
γ = Path-loss exponent

Then we can write:

$$\begin{aligned} RSL_A &\propto (d_A)^{-\gamma} \\ RSL_B &\propto (d_B)^{-\gamma} \end{aligned} \tag{4.1}$$

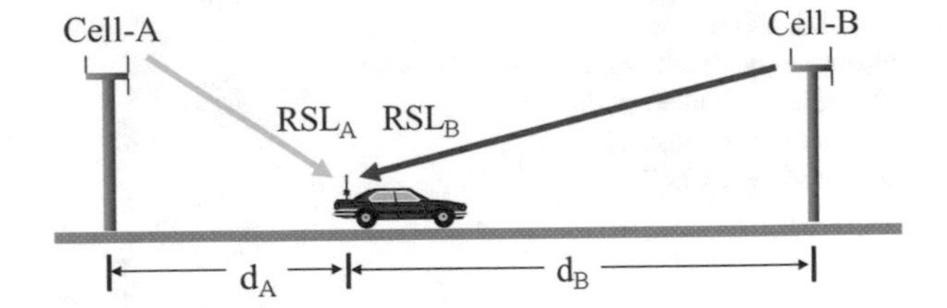

Fig. 4.7 Carrier-to-interference ratio (C/I) due to a single interferer

The ratio of the signal strengths at the mobile will be:

$$\frac{RSL_A}{RSL_B} = \left(\frac{d_A}{d_B}\right)^{-\gamma} = \left(\frac{d_B}{d_A}\right)^{\gamma} \tag{4.2}$$

RSL_A is the RF signal received from the serving cell. Therefore, this is the desired signal and we redefine this signal as the carrier signal C. We also assume that the mobile is at the cell edge from the serving Cell-A, which is the cell radius R. On the other hand, RSL_B is the undesired signal received from cells, and we redefine this signal as the interference signal I. The corresponding interference distance $d_B = D$; D being the reuse distance. Therefore, the above equation can be written as a carrier-to-interference ratio (C/I), due to a single interferer, as:

$$\frac{C}{I} = \left(\frac{D}{R}\right)^{\gamma} \tag{4.3}$$

In decibel, it can be written as:

$$\frac{C}{I}(dB) = 10\log\left(\frac{D}{R}\right)^{\gamma} \tag{4.4}$$

4.3.2 *C/I Due to Multiple Interferers*

In hexagonal cellular geometry, each hexagonal cell is surrounded by six hexagons. Therefore, in a mature cellular system, there can be six primary interferers as depicted in Fig. 4.8.

The total interference from all six interferers will be:

$$\begin{aligned} &6RSL_B \propto (d_B)^{-\gamma} \\ &or \\ &RSL_B \propto \frac{1}{6}(d_B)^{-\gamma} \end{aligned} \tag{4.5}$$

Therefore, the effective C/I is:

$$\frac{C}{I} = \frac{RSL_A}{RSL_B} = \frac{1}{6}\left(\frac{d_A}{d_B}\right)^{-\gamma} = \frac{1}{6}\left(\frac{d_B}{d_A}\right)^{\gamma} = \frac{1}{6}\left(\frac{D}{R}\right)^{\gamma} \tag{4.6}$$

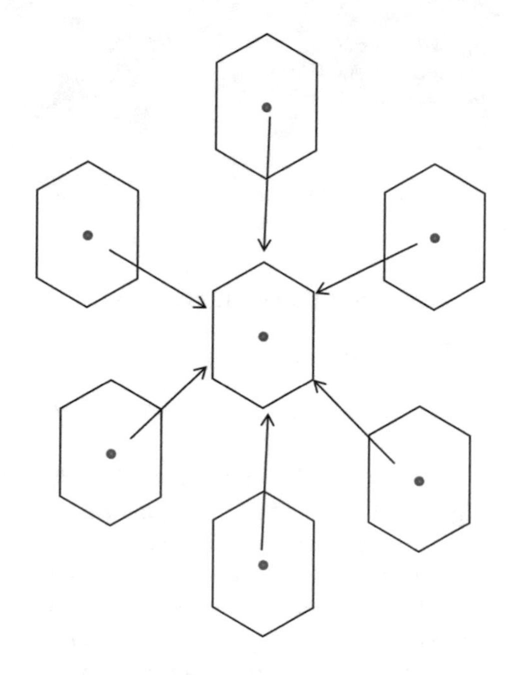

Fig. 4.8 C/I due to multiple interferers. Group of frequencies used in the center cell are reused in the surrounding six cells

and in decibel:

$$\frac{C}{I}(dB) = 10\log\left[\frac{1}{6}\left(\frac{D}{R}\right)^{\gamma}\right] \tag{4.7}$$

Therefore, by knowing the reuse distance, the C/I ratio can be determined. Or, by knowing the C/I requirement, the reuse distance can be determined in a given propagation environment. The reuse distance D can be determined from plane geometry, and the cell radius can be obtained from the propagation model. In general, C/I can be estimated as:

$$\frac{C}{I} = 10\log\left[\frac{1}{k}\left(\frac{D}{R}\right)^{\gamma}\right] \tag{4.8}$$

where

$$\begin{aligned} \mathrm{k} &= \text{Number of Co Channel Interferers} \\ &= 6(\text{OMNI}) \\ &= 3(\text{Sectorized}) \end{aligned}$$

γ = Propagation constant
D = Frequency reuse distance
R = Cell radius

The distance ratio (reuse distance) D/R is given by [4]:

$$\frac{D}{R} = \sqrt{3N} \tag{4.9}$$

where N is the number of cells per cluster.

4.4 Antenna Down Tilt

The 120° sectorization is achieved by dividing a cell into three sectors, while directional antennas are used in each sector. Thus, antenna configuration and their directivity play an important role in determining the C/I performances [5]. In order to illustrate this further, let us consider the diagram as shown in Fig. 4.9, where directional antennas are used for the present analysis. Antenna down tilt is also provided for additional isolation, which must be taken into account. The angle of down tilt is related to the cell radius and antenna height. This can be calculated by using plane geometry. It is given by the following equation:

$$\mathrm{R} = \mathrm{H}/(\tan\theta) \tag{4.10}$$

where

R = Cell radius
H = Antenna height
θ = Angle of antenna down tilt

These assumptions modify the C/I prediction equation as:

$$\frac{C}{I} = 10\log\left[\frac{1}{k}\left(\frac{D}{R}\right)^{\gamma}\right] + \Delta dB\ (\text{due to antenna down tilt}) \tag{4.11}$$

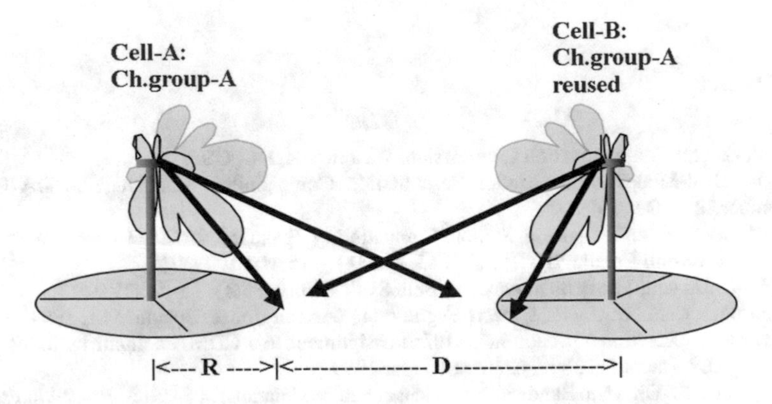

Fig. 4.9 Illustration of antenna down tilt

With $k = 6$, $\gamma = 4.84$, $D/R = 4.58$, and $\Delta dB \approx 6$ dB, we now obtain:

$$C/I \oplus 24.6 \text{ dB}. \quad (4.12)$$

Therefore, antenna down tilt improves C/I performance. The C/I performance can be further improved by using antenna having a narrow vertical beam width.

4.5 Optimization

Cellular communication has brought the world community closer than ever before; it is indispensable in even our everyday lives. Its use is increasing every day. However, it is unfortunate that the classical spectrum management techniques cannot keep pace with this because of customer growth and high-speed data communication. It is expected that this demand will increase exponentially. Therefore, additional measures are needed to support this demand. In an effort to address these issues, dynamic channel assignment is often used using the system intelligence at the NOC (Network Operation Center). It all begins with the classical frequency plan followed by optimization. The use of software-defined radio (SDR) has also been proposed to accomplish this [7].

4.6 Conclusions

A comprehensive, yet a concise overview of the fixed spectrum management technique is presented. Beginning with the FCC-allocated frequency band, we show how channels are created and allocated to different cells. Cell reuse plans are described next with C/I. The improvement of channel capacity by means of optimization is also described.

References

1. FCC, Federal Communication Commission, Washington, DC, USA
2. IS-54, Dual-Mode Mobile Station-Base Station Compatibility Standard EIA/TIA Project Number 2215, Dec 1989
3. IS-95, Mobile Station—Base Station Compatibility Standard for Dual Mode Wide Band Spread Spectrum Cellular Systems, TR 45, PN-3115, 15 Mar 1993
4. V.H. MacDonald, The cellular concept. Bell. Syst. Tech. J. **58**(1), 15–41 (1979)
5. S. Faruque, *Cellular Mobile Systems Engineering* (Artech House, Boston, MA, 1994)
6. S. Faruque, Directional pseudo-noise offset assignment in a CDMA cellular radio telephone system, U.S. Patent 5883889, Granted 16 Mar 1999
7. S. Faruque, M. Dhawan, Bandwidth efficient coded modulation for SDR, in *Proceedings of the SDR'10 Technical Conference and Product Exposition*, 2010, pp. 726–730

Chapter 5
Traffic Engineering

Abstract Traffic engineering is a branch of science that deals with provisioning of communication circuits in a given service area, for a given number of subscribers, with a given Grade of Service (GOS). It involves acquisition of population density per cell, translation of population density into traffic data (Erlang) per cell, computing the number of channels per cell using the Erlang table, and estimating the total number of cells in a given geographic area.

5.1 Introduction

Traffic engineering is a branch of science that deals with provisioning of communication circuits in a given service area, for a given number of subscribers, with a given Grade of Service (GOS) [1, 2]. It involves:

- Acquisition of population density per cell
- Translation of population density into traffic data (Erlang) per cell
- Computing the number of channels per cell using the Erlang table
- Estimating total number of cells in a given geographic area

Our primary objective in this chapter is to provide a basic understanding of traffic engineering and to present engineering aspects of cell site provisioning. Since 5G cellular standard is still evolving, we will review the key concepts and underlying principles and address the salient features of traffic engineering. This will form the foundation of traffic engineering.

5.2 Over-provisioning and Under-provisioning

Traffic engineering is also a process of revenue generation, which is crucial for service providers. An over-provisioned system, as shown in Fig. 5.1a, guarantees 100% system availability, but it is not cost-effective. On the other hand, an

S. Faruque, *Radio Frequency Cell Site Engineering Made Easy*, SpringerBriefs in Electrical and Computer Engineering, https://doi.org/10.1007/978-3-319-99615-8_5

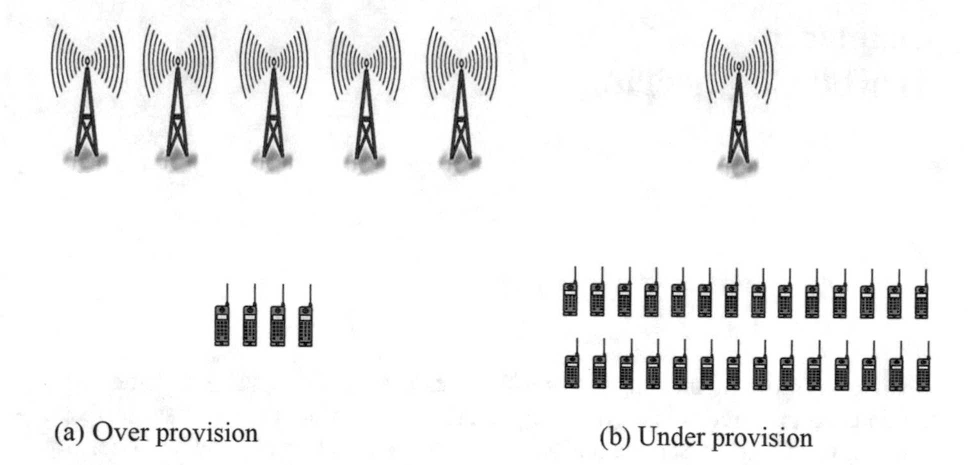

Fig. 5.1 Conceptual representation of over-provision (**a**) and under-provision (**b**). A good compromise between these two is the art of traffic engineering

under-provisioned system (Fig. 5.1b), having too many subscribers per channel, is responsible for call blocking, denying services to many subscribers. A good compromise between these two is the art of traffic engineering.

5.3 Traffic Characteristics

All traffic engineering is based on the average busy hour traffic during a day. This busy hour is generally consistent and predictable. A typical distribution pattern of cellular traffic is given in Fig. 5.2 [1]. It has two peaks: before noon and before evening. These peaks are known as busy hour traffic. Traffic is generally low during the night and rises rapidly in the morning when offices, shops, and factories open for business. Traffic intensity gradually goes down during lunch time and again picks up in the afternoon. These variations are commonly known as hourly variations.

Figure 5.2 also indicates higher average intensities on business days and lower traffic activities during weekends and holidays. These variations are known as daily variations.

There are other variations as well, as described here:

- *Call holding time variations*: Call holding time is the average duration of a call. It varies according to types of subscriber (i.e., business, private, etc.). Typical call holding time varies between 120 and 180 s. This is an important parameter in estimating the traffic, as we shall see later.
- *Seasonal variations*: Pre-Christmas, Pre-Easter, etc. It may be necessary to provision more circuits for those special days.
- *Long-term variations*: Gradual subscriber growth over a period of years. These variations have to be taken into account for long-term growth and system planning.

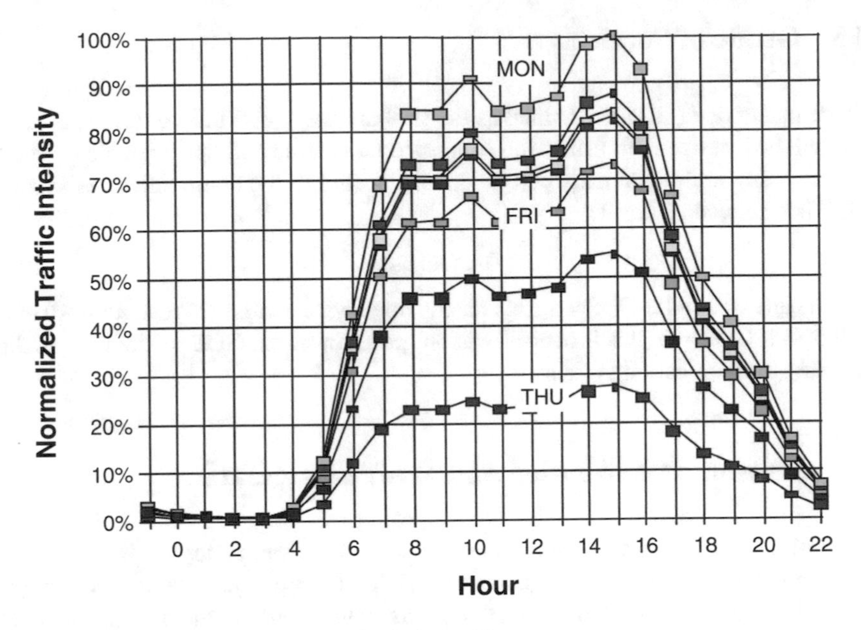

Fig. 5.2 A typical distribution pattern of cellular traffic

5.4 Intensity and Units of Traffic

Traffic intensity is measured in Erlang, where 1 Erlang = one circuit in use for 1 h (3600 s), named after the Danish mathematician, A.K. Erlang, founder of the theory of telephone traffic. Traffic intensity is also measured in CCS per hour (Circuit Centum Second), where 1 CCS = one circuit in use for 100 s. The relationship between Erlang and CCS can be defined as follows:

- 1 Erlang = 1 ckt. In-use for 3600 s
- 1 CCS = 1 ckt. In-use for 100 s

Therefore, Erlang/CCS = 36 or 1 Erl. = 36 CCS. These traffic units are defined as:

$$\text{Erlang} = \frac{\text{Number of calls} \times \text{Average call holding time (s)}}{3600} \tag{5.1}$$

$$\text{CCS} = \frac{\text{Number of calls} \times \text{Average call holding time (s)}}{100} \tag{5.2}$$

Either Erlang or CCS can be used to determine the total number of channels required to meet the service objectives for a given blocking rate. Number of calls is the total number of calls that can be handled for a given Erlang or CCS.

5.5 Grade of Service

Grade of Service (GOS) is defined as the probability of call failure. It means that a call will be lost due to transmission congestion, i.e., when all the available channels are busy, any additional calls will be denied access to the communication system. GOS lies between 0 and 1:

$$0 < \text{GOS} < 1 \tag{5.3}$$

All calls will fail if GOS = 1 (this means no service at all, zero revenue). All calls will pass if GOS = 0 (this is over-provision, poor revenue). GOS = 0.02 is typical in cellular communication systems.

5.6 Poisson's Distribution and Traffic Calculations

The Poisson distribution [2–5] is a statistical process, which applies to a sequence of events that takes place at regular intervals of time or throughout a continuous interval of time. The Poisson distribution has many important applications, where we may be interested in the number of customers arriving for service at a pit stop or the number of airplanes arriving at an airport or the number of phone calls arriving at a switch. The mathematical model, which describes many situations like these, is the Poisson distribution. Let N be the total number of trunks (channels) and T be the offered traffic in Erlang (or CCS). Then the following Poisson distribution will give the probability of all the channels being busy:

$$P(N,T) = \frac{T^N e^{-T}}{N!} \tag{5.4}$$

where P(N, T) is the blocking rate or GOS (Grade of Service). Thus, for a given traffic capacity and blocking rate, the number of radios can be calculated. Table 5.1 provides a list of offered traffic in Erlang as a function of blocking probability. Additional traffic data for GOS other than 2% may be obtained from elsewhere [4, 5]. Note that the most often used table for telephony is the Erlang B table. It assumes that blocked calls are cleared and the caller tries again later. There are other tables such as Erlang-C, which assumes blocked calls retry until the call is established.

Figure 5.3 shows the relationship between Erlang and the number of channels for GOS = 2% which can be used for a quick estimation of traffic for 2% blocking rate.

Example *Given*: 20 subscribers, each of which generates 0.1 Erlang/h and 4 available channels.

What is the probability that all four channels will be busy?
Answer:
N = 4
E = 0.1 × 20 = 2

Table 5.1 Erlang B table with GOS = 2% (most frequently used for cellular applications)

#Trunks	Erlang	#Trunks	Erlang	#Trunks	Erlang	#Trunks	Erlang	#Trunks	Erlang	#Trunks	Erlang	#Trunks	Erlang	#Trunks	Erlang
1	0.0204	26	18.4	51	41.2	76	64.9	100	88	150	136.8	200	186.2	250	235.8
2	0.223	27	19.3	52	42.1	77	65.8	102	89.9	152	138.8	202	188.1	300	285.7
3	0.602	28	20.2	53	43.1	78	66.8	104	91.9	154	140.7	204	190.1	350	335.7
4	0.109	29	21	54	44	79	67.7	106	93.8	156	142.7	206	192.1	400	385.9
5	1.66	30	21.9	55	44.9	80	68.7	108	95.7	158	144.7	208	194.1	450	436.1
6	2.28	31	22.8	56	45.9	81	69.6	110	97.7	160	146.6	210	196.1	500	486.4
7	2.94	32	23.7	57	46.8	82	70.6	112	99.6	162	148.6	212	198.1	600	587.2
8	3.63	33	24.6	58	47.8	83	71.6	114	101.6	164	150.6	214	200	700	688.2
9	4.34	34	25.5	59	48.7	84	72.5	116	103.5	166	152.6	216	202	800	789.3
10	5.08	35	26.4	60	49.6	85	73.5	118	105.5	168	154.5	218	204	900	890.6
11	5.84	36	27.3	61	50.6	86	74.5	120	107.4	170	156.5	220	206	1000	999.1
12	6.61	37	28.3	62	51.5	87	75.4	122	109.4	172	158.5	222	208	1100	1093
13	7.4	38	29.2	63	52.5	88	76.4	124	111.3	174	160.4	224	210		
14	8.2	39	30.1	64	53.4	89	77.3	126	113.3	176	162.4	226	212		
15	9.01	40	31	65	54.4	90	78.3	128	115.2	178	164.4	228	213.9		
16	9.83	41	31.9	66	55.3	91	79.3	130	117.2	180	166.4	230	215.9		
17	10.7	42	32.8	67	56.3	92	80.2	132	119.1	182	168.3	232	217.9		
18	11.5	43	33.8	68	57.2	93	81.2	134	121.1	184	170.3	234	219.9		
19	12.3	44	34.7	69	58.2	94	82.2	136	123.1	186	172.4	236	221.9		
20	13.2	45	35.6	70	59.1	95	83.1	138	125	188	174.3	238	223.9		
21	14	46	36.5	71	60.1	96	84.1	140	127	190	176.3	240	225.9		
22	14.9	47	37.5	72	61	97	85.1	142	128.9	192	178.2	242	227.9		
23	15.8	48	38.4	73	62	98	86	144	130.9	194	180.2	244	229.9		
24	16.6	49	39.3	74	62.9	99	87	146	132.9	196	182.2	246	231.8		
25	17.5	50	40.3	75	63.9	100	88	148	134.8	198	184.2	248	233.8		

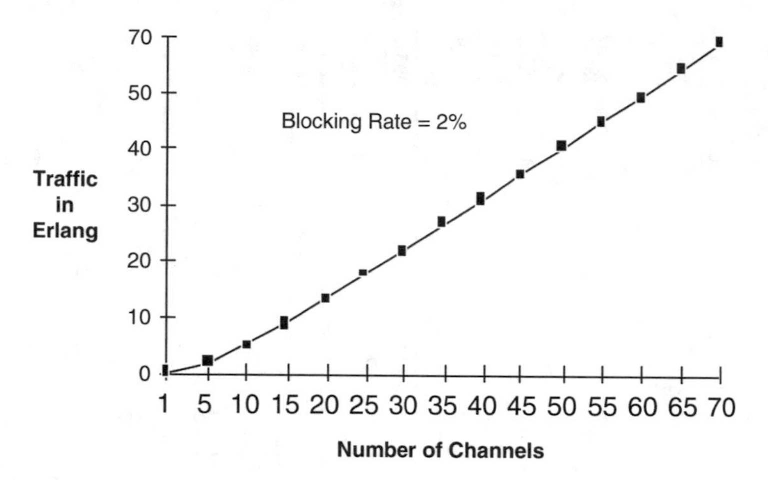

Fig. 5.3 Relationship between Erlang and number of channels

$$P(4,2) = \frac{2^4 e^{-2}}{4!} = 0.09$$

That is, there is a 9% probability of four channels being busy or GOS = 0.09.

Example Given the following parameters, determine the total number of channels. Assume number of calls expected = 1000, average call holding time = 120, and GOS = 2%.

Average call holding time = 120 s

Answer:

Total traffic in Erlangs = 1000 × 120/3600 ≈ 33 Erlangs

From the Erlang B table or from the graph of Fig. 5.4, we find that 33 Erlang @2% GOS represents approximately 42 channels. Therefore, for all practical purposes, an OMNI site having 48 channels is a good compromise.

5.7 Trunking Efficiency

Trunking efficiency (channel utilization efficiency) is also known as a measure of base station efficiency. It is determined by the amount of traffic per channel, defined as:

$$\text{Efficiency}\ (\%) = \frac{\text{Traffic in Erlangs}}{\text{Number of Channels}} \times 100 \tag{5.5}$$

From the Erlang table with GOS = 2%, we obtain the graph of Fig. 5.4, which shows the relationship between efficiency and capacity. We see that for a given GOS, the efficiency increases as the number of trunks (voice circuits) increases.

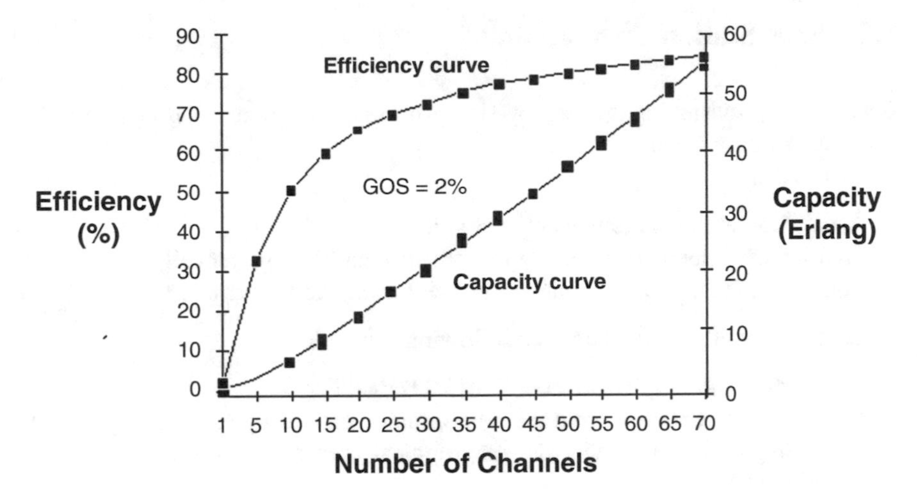

Fig. 5.4 Trunking efficiency curve

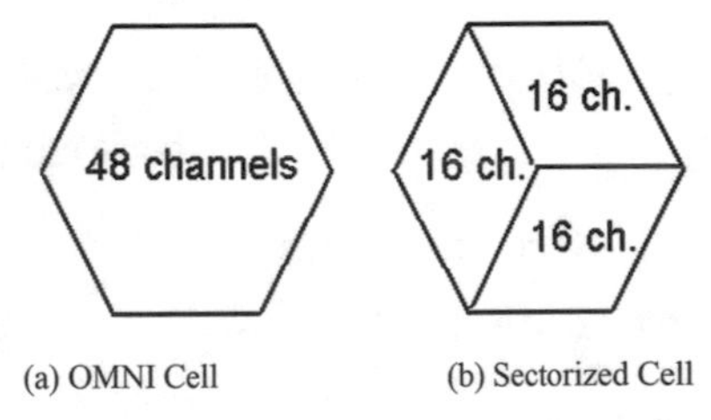

Fig. 5.5 Illustration of trunking efficiency. (**a**) OMNI Cell. (**b**) Sectorized Cell

A cell site having <15 voice circuits (channels) is generally inefficient, less cost-effective and generates poor revenue.

To illustrate this further, we consider an OMNI cell having 48 channels, which are then sectorized into three as shown in Fig. 5.5. Although the total number of channels in the sectorized cell remains the same, the traffic capacity of the OMNI cell is higher than the sectorized cell due to trunking efficiency.

Example *For the Omni Site*:

Total number of channels available = 48
GOS = 0.02
Then the traffic intensity in Erlang will be = 38.4 Erl.
Therefore, trunking efficiency = 38.4/48 = 0.8 (80%)
For the Sectorized Site:
The available 48 channels are divided into three groups, 16 channels per group.
Number of channels/group = 16
GOS = 0.02
Then the traffic intensity/group in Erl. = 9.83
Total traffic in three sectors (3 × 9.83) = 29.49 Erl.
Therefore trunking efficiency = 29.49/48 = 0.514 (51.4%)

5.8 Base Station Provisioning

Base station provisioning is a step-by-step process of assigning a certain number of channels per base station.

It involves:

- Acquisition of population density per cell
- Translation of population density into traffic data (Erlang) per cell
- Computing the number of channels per cell using the Erlang table

These are briefly described in the following subsections:

Step 1: Acquisition of Population Density Per Cell

This can be obtained from the demographic data or by using Google Earth. For example, a cell in a given geographical area captures 1000 subscribers (see Fig. 5.6).

Step 2: Traffic Count in Erlang

Traffic count in Erlang is derived from the following equation:

$$\text{Total traffic in Erlangs} = \text{Total Number of Subscribers} \times \text{Call Holding Time} / 3600$$
$$= 1000 \times 120 / 3600 \approx 33 \text{ Erlangs}$$

This is shown in Fig. 5.7.

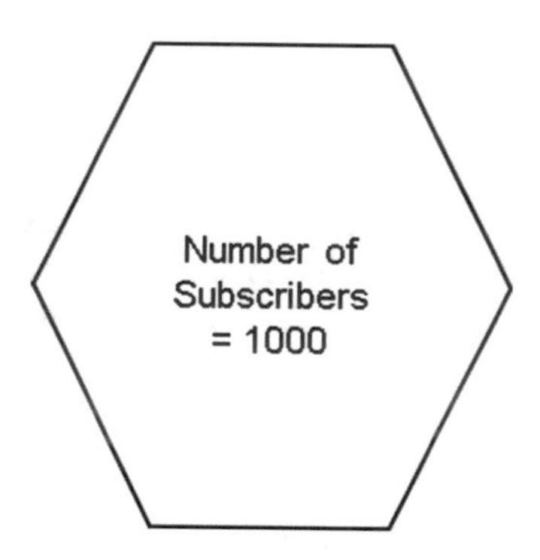

Fig. 5.6 Acquisition of population density in a given cell, in a given geographical area. There are 1000 possible subscribers in this cell

Fig. 5.7 Traffic in Erlang 1

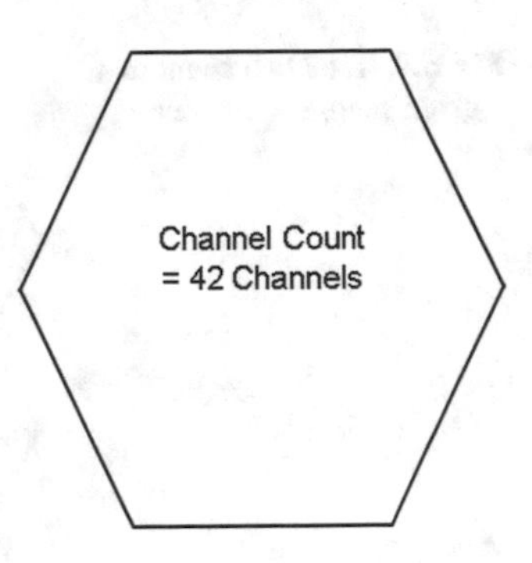

Fig. 5.8 Channel count

Step 3: Channel Count

Once the traffic count is determined in terms of Erlang, the number of channels can be determined by using the Erlang B table. Because of the high cost of base station radios, it is customary to use a high GOS. A GOS of 2% is generally used for cellular base stations. This is shown in Fig. 5.8.

5.9 Cell Count

Once the number of channels per cell is determined from the Erlang B table, the total cell count in a given geographical area can be obtained as shown in Fig. 5.9. Notice that the channel count is different for different cell. This is due to the fact that the traffic distribution pattern is not uniform. Therefore channel usage varies from cell to cell. This brings the traffic engineering process to an end.

5.10 Conclusions

Traffic engineering is a branch of science that deals with provisioning of communication circuits in a given service area, for a given number of subscribers, with a given Grade of Service (GOS). It involves acquisition of population density per cell, translation of population density into traffic data (Erlang) per cell, computing the number of channels per cell using the Erlang table, and estimating the total number of cells in a given geographic area. This chapter provides the key concepts and underlying principles of traffic engineering.

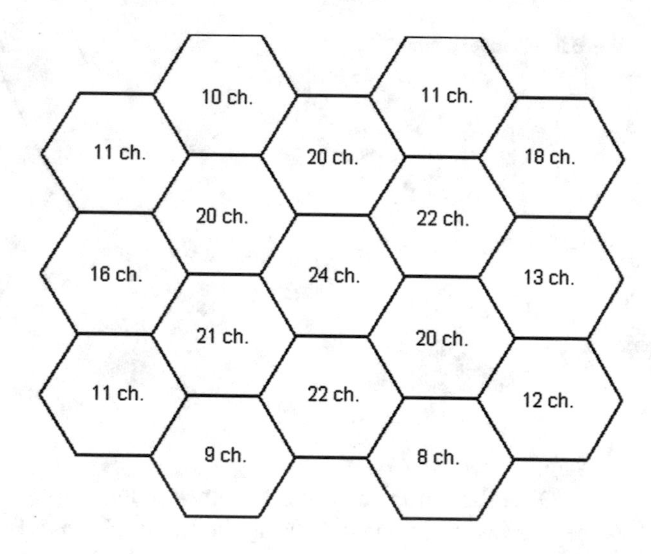

Fig. 5.9 Total cell count in a given geographical area

References

1. S. Baxter, S. Faruque, *RF Engineering Seminar 1000, An Internal RF Engineering Course* (Northern Telecom, Richardson, TX, 1994)
2. S. Faruque, *Cellular Mobile Systems Engineering* (Artech House Inc., Norwood, MA, 1996). ISBN: 0-89006-518-7
3. E. Kreyszig, *Advanced Engineering Mathematics*, 5th edn. (John Wiley & Sons, New York, 1983)
4. I. Miller, J.E. Freund, *Probability and Statistics for Engineers* (Prentice-Hall Inc., Englewood Cliffs, NJ, 1955)
5. W.C.Y. Lee, *Mobile Cellular Telecommunications Systems* (McGraw-Hill Book Company, New York, 1989)

Chapter 6
Cell Site Optimization

Abstract Cell site optimization involves live air data collection as a function of distance and statistical analysis of the data for adjusting RF coverage footprints. This concluding chapter shows how these tools are used to collect live air data and perform statistical analysis to optimize the cell site.

6.1 Introduction

Cell site optimization involves live air data collection as a function of distance and statistical analysis of the data for adjusting RF coverage footprints. Since multipath propagation is fuzzy owing to numerous RF barriers, uneven terrain, hills, trees, buildings, etc., there is a large variation of received signal level (RSL) at the receiver [1–5].

Figure 6.1 shows an example to illustrate this scenario. Here, the received signal level (RSL) is measured in dBm and the distance is measured in km. Notice that the signal strength decays logarithmically as a function of distance where the distance is plotted in the linear scale. Also notice that we have plotted a solid line, which is known as the regression line or the best fit. The regression line has a special significance since 50% data are above the regression line and 50% data are below the regression line. Furthermore, at a given distance, the distribution of data has a shape known as Gaussian (bell shaped). We shall discuss these points along with their attributes later in this chapter. Today, a large number of PC-based data collection tools are commercially available. These tools are capable of importing measurement data and of generating statistical outputs such as mean error, standard deviation, max., min., etc.

These data collection tools are capable of importing measurement data and generating statistical outputs such as mean error, standard deviation, max., min., etc.

In this chapter, we will see how these tools are used to collect live air data and perform statistical analysis to optimize the cell site. Let's take a closer look!

S. Faruque, *Radio Frequency Cell Site Engineering Made Easy*, SpringerBriefs in Electrical and Computer Engineering, https://doi.org/10.1007/978-3-319-99615-8_6

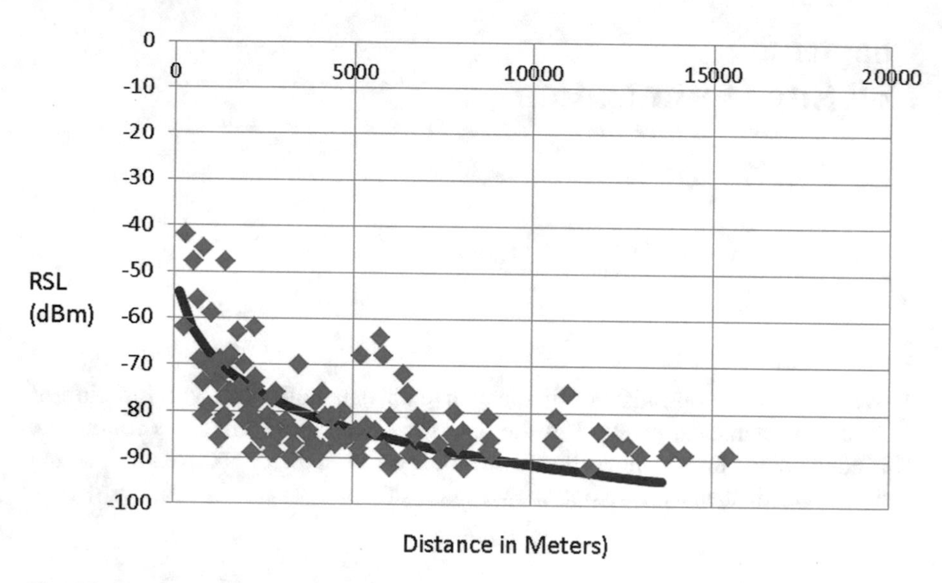

Fig. 6.1 Received signal strength as a function of distance in semilogarithmic scale

6.2 Live Air Data Collection

Figure 6.2 shows the basic concept of RF data collection technique. It is PC based and uses a cellular radio, MS Excel, a GIS (Geographic Information Services) software, and a GPS (Global Positioning System) receiver. The GPS receiver is used to collect the coordinates (longitude and latitude) of each sampling point. The outcome is a pair of long/lat corresponding to each RSL value. Since the cell site location is fixed and has a unique long/lat value, the distance of each sampling point with respect to the cell site is readily available as an output.

Table 6.1 shows an output file, which was obtained by means of drive test. Notice that the received signal level (RSL) is measured in dBm as a function of distance, where the distance is in meters. We can now perform statistical analysis to find the following parameters:

- Mean
- Standard deviation
- Minimum RSL value
- Maximum RSL value and
- The regression line

The above statistical parameters were calculated in Excel and also presented in Table 6.1 at the end.

The above data was collected from a cell site in a typical urban environment. Figure 6.3 shows the received signal level (RSL) as a function of distance in a semi-logarithmic scale. RSL is measured in dBm and the distance is measured in meters. Here, we see that the signal strength decays logarithmically as a function of distance where the distance is plotted in the linear scale. The rate of decay depends on the propagation environment.

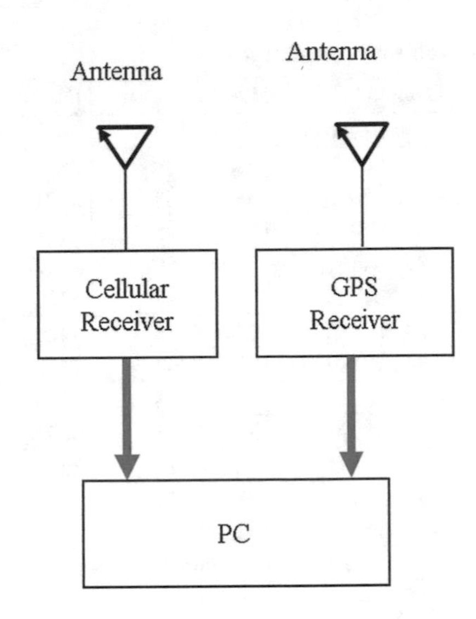

Fig. 6.2 Drive test and data collection technique. The received signal level (RSL) is measured as a function of distance

Table 6.1 Output file obtained from drive test

Longitude	Latitude	Distance_(m)	RSL(dBm)_900 MHz	
−84.409384	33.181533	269	−62	
−84.39526	33.181423	1265	−86	
−84.402058	33.186028	666	−56	
−84.382661	33.188661	2436	−86	
−84.392245	33.190038	1658	−66	
−84.392885	33.199833	2289	−68	
−84.465622	33.888264	13,041	−89	
−84.364382	33.848223	6691	−85	
−84.368541	33.824409	5268	−86	
−84.394605	33.198652	2080	−80	
−84.410685	33.196834	1553	−66	
−84.366262	33.813226	4398	−81	
−84.361534	33.814066	4660	−80	
−84.416639	33.166681	966	−69	
−84.420696	33.163503	1569	−66	
−84.413636	33.80609	2611	−81	
−84.439221	33.815618	4463	−85	
−84.32865	33.828839	8818	−81	
−84.380246	33.838196	6564	−66	
−84.341511	33.831954	8118	−84	
−84.394041	33.186416	1403	−81	
−84.355556	33.82698	6860	−81	
−84.439033	33.803118	3464	−60	
−84.421633	33.801042	2232	−62	
−84.346082	33.81868	6698	−85	
−84.434166	33.8399	6636	−89	
−84.406699	33.808521	2645	−82	

(continued)

Table 6.1 (continued)

Longitude	Latitude	Distance_(m)	RSL(dBm)_900 MHz	
−84.431156	33.812693	3666	−84	
−84.406158	33.809509	2852	−66	
−84.335608	33.835658	8821	−88	
−84.291993	33.885625	15,481	−89	
−84.46436	33.880669	11,864	−84	
−84.394885	33.196343	1961	−60	
−84.383886	33.826268	5220	−90	
−84.310189	33.859344	12,264	−86	
−84.406229	33.196068	1360	−82	
−84.34899	33.822011	6891	−90	
−84.426492	33.804313	2663	−82	
−84.426449	33.196391	2165	−89	
−84.414065	33.808396	2656	−66	
−84.419483	33.194658	1522	−69	
−84.361092	33.819365	5230	−68	
−84.399822	33.816499	3619	−85	
−84.422323	33.180823	1260	−63	
−84.406662	33.18591	293	−42	
−84.40196	33.169116	826	−81	
−84.415423	33.800848	1968	−82	
−84.411428	33.188033	512	−48	
−84.414492	33.19446	1269	−69	
−84.416928	33.18921	1005	−61	
−84.388593	33.802899	2699	−89	
−84.412358	33.154238	3312	−84	
−84.402105	33.164061	2286	−84	
−84.409161	33.199691	1663	−63	
−84.36565	33.820403	5640	−84	
−84.433684	33.806493	3356	−86	
−84.468485	33.83561	6866	−80	
−84.480314	33.828859	8153	−92	
−84.426616	33.836318	6049	−92	
−84.4168	33.182528	631	−69	
−84.424343	33.185529	1413	−66	
−84.438201	33.810289	3948	−68	
−84.399288	33.193661	1401	−48	
−84.313666	33.886651	14,261	−89	
−84.410956	33.813269	3266	−90	
−84.399264	33.195825	1584	−68	
−84.329951	33.815803	8003	−86	
−84.399336	33.188809	1029	−59	
−84.364188	33.82932	6446	−62	
−84.36128	33.801496	3936	−86	
−84.360836	33.822189	6066	−81	
−84.366214	33.823361	5680	−64	
−84.383866	33.896086	12,664	−86	

(continued)

Table 6.1 (continued)

Longitude	Latitude	Distance_(m)	RSL(dBm)_900 MHz	
−84.406141	33.908496	13,832	−88	
−84.425185	33.814958	3645	−89	
−84.361419	33.904251	13,665	−89	
−84.406483	33.191069	803	−45	
−84.409319	33.83035	5156	−88	
−84.411533	33.839084	6135	−90	
−84.382223	33.85451	8218	−86	
−84.414984	33.164108	1222	−64	
−84.481259	33.863226	11,008	−66	
−84.469531	33.866018	10,682	−81	
−84.392566	33.804818	2661	−82	
−84.408506	33.824264	4469	−81	
−84.384223	33.180869	2268	−82	
−84.364806	33.818304	4925	−85	
−84.364025	33.841106	6096	−82	
−84.442113	33.191291	3116	−82	
−84.454986	33.199635	4536	−86	
−84.36486	33.816318	5385	−83	
−84.362066	33.806133	4236	−81	
−84.391312	33.886535	11,618	−92	
−84.366962	33.8106	4160	−66	
−84.368686	33.812433	4813	−86	
−84.384083	33.820446	4626	−85	
−84.485096	33.856248	10,593	−86	
−84.349366	33.846486	8895	−89	
−84.361439	33.826041	5849	−68	
−84.343123	33.824095	6460	−86	
−84.382661	33.832415	5868	−88	
−84.425408	33.163462	2614	−86	
−84.388039	33.16346	2945	−85	
−84.406436	33.166603	822	−64	
−84.431689	33.199861	2640	−89	
−84.388655	33.184356	1826	−65	
−84.388336	33.168003	1984	−66	
−84.388363	33.162162	2263	−65	
−84.388658	33.19639	2285	−63	
−84.364855	33.812618	5103	−84	
−84.330568	33.831609	8883	−86	
−84.410364	33.804906	2339	−85	
−84.381869	33.814389	4166	−88	
−84.403161	33.819389	3966	−90	
−84.429242	33.849496	6505	−90	
		Mean=	−69.86826086	dBm
		Stdev=	10.06963613	dB
		Max=	−42	dBm
		Min=	−92	dBm

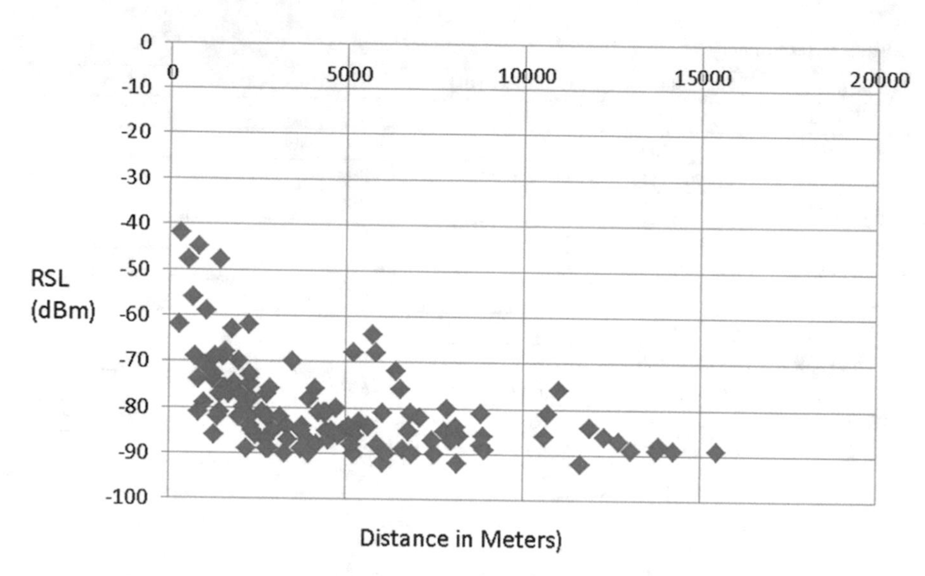

Fig. 6.3 The received signal level (RSL) as a function of distance in a semilogarithmic scale. RSL is measured in dBm and the distance is measured in meters

Our analysis indicates that the cell site exhibits following performance characteristics:

Mean RSL = −69.8 dBm
Standard Dev. = 10.08 dB
Max. RSL = −42 dBm
Mi. RSL = −92 dBm

These values are typical in urban environment and the cell site is healthy. The 10 dB standard deviation has a special significance in designing reliable cell sites, which we shall see next.

6.3 Statistical Analysis and Optimization

Statistics is the study of the collection, organization, analysis, interpretation, and presentation of data [6–11]. For radio-frequency (RF) engineering, it involves live air data collection as a function of distance. Since multipath propagation is fuzzy owing to numerous RF barriers, uneven terrain, hills, trees, buildings, etc., there is a large variation of received signal level (RSL) at the receiver [5].

Now we consider a set of random variables RSL_i having n sample values where i = 1, 2, …, n. The distribution or the density of such a set of random numbers is generally approximated by a continuous curve known as normal distribution. The equation that describes a normal distribution is given by [9, 10]:

$$f(RSL) = \frac{1}{\sigma\sqrt{2\pi}} \exp\left(-0.5\left[\left(\frac{RSL - \overline{RSL}}{\sigma}\right)\right]^2\right) \tag{6.1}$$

where the mean is given by:

$$\overline{RSL} = \frac{\mathrm{RSL}_1 + \mathrm{RSL}_2 + \ldots + \mathrm{RSL}_n}{n} \tag{6.2}$$

and the variance is given by:

$$\sigma^2 = \frac{\left(\mathrm{RSL}_1 - \overline{RSL}\right)^2 + \left(\mathrm{RSL}_2 - \overline{RSL}\right)^2 + \ldots + \left(\mathrm{RSL}_n - \overline{RSL}\right)^2}{n-1} \tag{6.3}$$

σ being the standard deviation.

The curve of Fig. 6.4 is also known as the Gaussian distribution or a bell-shaped curve which is symmetric with respect to the mean whose peak at $\overline{RSL} = 0$ increases as σ decreases.

Figure 6.5 shows the distribution curve for $\overline{RSL} \uparrow 0$. We notice that for a positive mean, the curve has the same shape but is shifted to the right and, for a negative mean, it is shifted to the left. This illustrates the fact that the variance is the average dispersion from the mean.

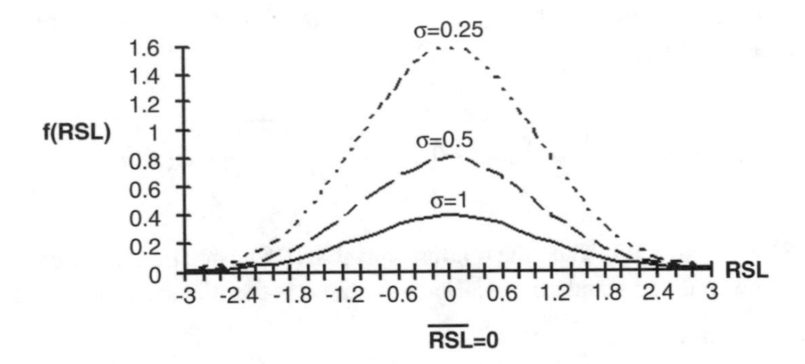

Fig. 6.4 Normal distribution with zero mean (RSL = 0) and variable standard deviation

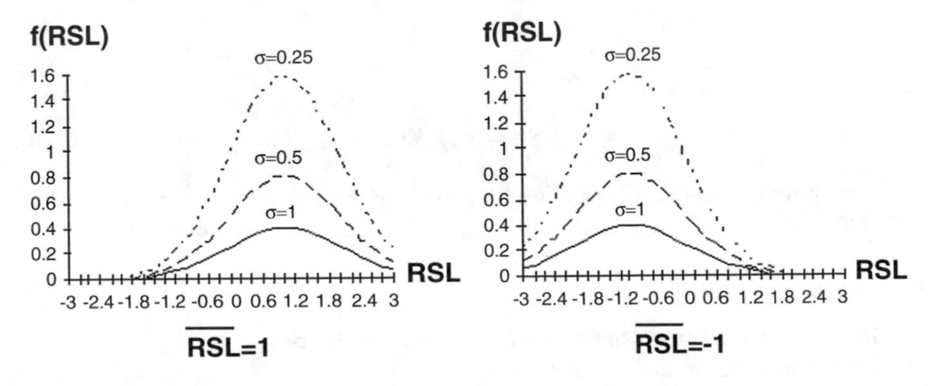

Fig. 6.5 Normal distribution with RSL = 1 and RSL = −1, s = variable

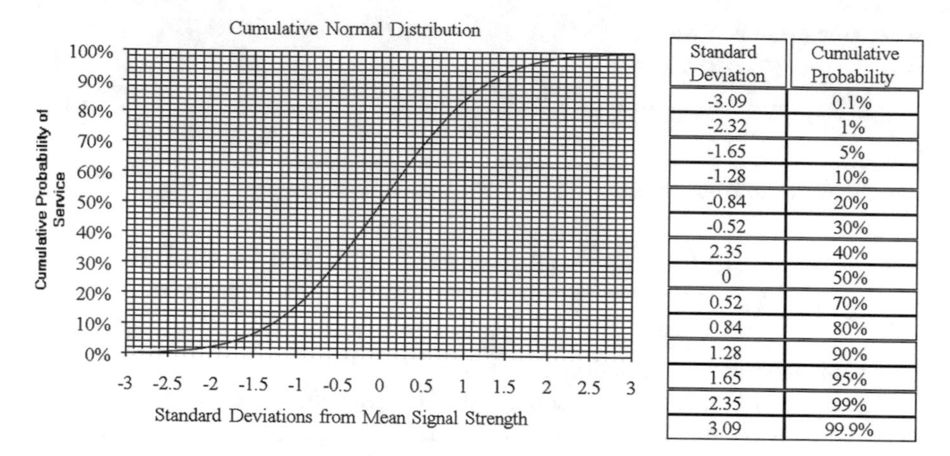

Standard Deviation	Cumulative Probability
-3.09	0.1%
-2.32	1%
-1.65	5%
-1.28	10%
-0.84	20%
-0.52	30%
2.35	40%
0	50%
0.52	70%
0.84	80%
1.28	90%
1.65	95%
2.35	99%
3.09	99.9%

Fig. 6.6 Cumulative probability distribution as a function of normalized standard deviation(z)

The probability density function is generally obtained from the standard table called standard normal distribution or by means of a curve called cumulative distribution function as shown in Fig. 6.6. Both are based on the following probability distribution function:

$$\mathrm{F}(z)=\frac{1}{\sigma\sqrt{2\pi}}\int_{-\infty}^{z}\exp\left(-0.5\left[\left(\frac{RSL-\overline{RSL}}{\sigma}\right)\right]\right)^{2}d(RSL) \tag{6.4}$$

with $\overline{RSL}=0$ and $\sigma=1$. Then the random variable (RSL) can be estimated from the following normalized standard deviation z where σ is the measured standard deviation [5]:

$$\mathrm{z}=\left(\frac{RSL-\overline{RSL}}{\sigma}\right) \tag{6.5}$$

or

$$\mathrm{RSL}=\sigma \mathrm{z}+\overline{RSL} \tag{6.6}$$

The above cumulative distribution function (Fig. 6.6) along with Eq. (6.6) forms the basis of cell site optimization technique. To illustrate the concept, here is an example:

Given:

- Desired received signal at the cell edge: RSL = −80 dBm
- Measured standard deviation: $\sigma = 8$ dB
- Required confidence level is 80%

Find:

- The minimum received signal level at the cell edge RSL to satisfy the above requirements

Solution:
For 80% confidence level, using the curve in Fig. 6.6, we have:

$$z(0.8) = 0.842.$$

Therefore, the minimum RSL can be computed as:

$$\begin{aligned} \text{RSL} &= \sigma z + \overline{RSL} \\ &= (8 \times 0.842) - 80 \text{ dBm} \\ &= -63.26 \text{ dBm}. \end{aligned}$$

- This is the optimized signal strength at the cell edge, which is stronger than the specified RSL.
- It ensures that 80% of the data will fall within the interval $-\sigma$ and $+\sigma$, i.e., within +8 dB.
- This interval is called the confidence interval and the probability (80%) is called the confidence level.

6.4 Conclusions

- Drive test, live air data collection, and data analysis techniques were presented.
- Reviewed statistical analysis and showed that random data such as received signal level (RSL) can be predicted for cell site optimization with confidence.

References

1. M. Hata, Empirical formula for propagation loss in land mobile radio services. IEEE Trans. Veh. Technol. **VT-29**, 317–326 (1980)
2. J. Walfisch et al., A theoretical model of UHF propagation in urban environments. IEEE Trans. Antennas Propag, AP-38, 1788–1796 (1988)
3. IS-54, Dual-Mode Mobile Station—Base Station Compatibility Standard, Electronic Industries Association Engineering Department, PN-2216, Dec 1989
4. H.H. Xia et al., Radio propagation characteristics for line of sight micro cellular and personal communications. IEEE Trans. Antennas Propag. **41**(10), 1439–1447 (1993)
5. S. Faruque, *Cellular Mobile Systems Engineering* (Artech House Inc., Norwood, MA, 1996). ISBN: 0-89006-518-7
6. I. Miller, J.E. Freund, *Probability and Statistics for Engineers* (Prentice-Hall Inc., Englewood Cliffs, NJ, 1977)

7. D.A. Freedman, *Statistical Models: Theory and Practice* (Cambridge University Press, New York, 2005)
8. G.U. Yule, On the theory of correlation. J. R. Stat. Soc. (Blackwell Publishing) **60**(4), 812–854 (1897). https://doi.org/10.2307/2979746. JSTOR 2979746
9. J.S. Armstrong, Illusions in regression analysis. Int. J. Forecasting **28**(3), 689 (2012). Forthcoming. https://doi.org/10.1016/j.ijforecast.2012.02.001
10. R.A. Fisher, The goodness of fit of regression formulae, and the distribution of regression coefficients. J. R. Stat. Soc. (Blackwell Publishing) **85**(4), 597–612 (1922). https://doi.org/10.2307/2341124. JSTOR 2341124
11. C. Tofallis, Least squares percentage regression. J. Mod. Appl. Stat. Methods **7**, 526–534 (2009). https://doi.org/10.2139/ssrn.1406472